国家级实验教学示范中心系列规划教材
普通高等院校机械类"十三五"规划实验教材

# 材料力学实验

# Experiments of Mechanics of Materials

## （双语版）

主　编　魏义敏
副主编　乐忠萍　周　迅
主　审　李剑敏

华中科技大学出版社
中国·武汉

# 内 容 简 介

本书是浙江理工大学机械设计制造系力学课程组根据教育部高等学校工科基础课程教学指导委员会公布的《材料力学课程教学基本要求(A类)》(2019 版)编写的。为配合全英文材料力学实验课程的教学需要,本书采用了中英文对照的形式且中英文部分内容基本相同。全书内容涵盖了高等学校机械类本科专业材料力学实验教学的内容,共分为两个部分(材料的力学性能实验和电测法应力分析实验),包含 9 个实验。

本书可作为高等学校本科机械类各专业的材料力学实验课程教材,也可在各高校机械类专业国际化全英文教学中使用。

**图书在版编目(CIP)数据**

材料力学实验:双语版/魏义敏主编.—武汉:华中科技大学出版社,2020.11
ISBN 978-7-5680-0961-4

Ⅰ.①材…　Ⅱ.①魏…　Ⅲ.①材料力学-实验-高等学校-教材　Ⅳ.①TB301-33

中国版本图书馆 CIP 数据核字(2020)第 215296 号

---

**材料力学实验(双语版)**
Cailiao Lixue Shiyan (Shuangyu Ban)

魏义敏　主编

---

策划编辑:万亚军
责任编辑:程　青
封面设计:原色设计
责任监印:周治超
出版发行:华中科技大学出版社(中国·武汉)　　电话:(027)81321913
　　　　　武汉市东湖新技术开发区华工科技园　　邮编:430223
录　　排:武汉市洪山区佳年华文印部
印　　刷:武汉科源印刷设计有限公司
开　　本:787mm×1092mm　1/16
印　　张:8.25
字　　数:212 千字
版　　次:2020 年 11 月第 1 版第 1 次印刷
定　　价:28.00 元

---

国家级实验教学示范中心系列规划教材
普通高等院校机械类"十三五"规划实验教材
# 编 委 会

丛书主编　吴昌林　华中科技大学

丛书编委（按姓氏拼音顺序排列）

邓宗全　哈尔滨工业大学

葛培琪　山东大学

何玉林　重庆大学

黄　平　华南理工大学

孔建益　武汉科技大学

蒙艳玫　广西大学

芮执元　兰州理工大学

孙根正　西北工业大学

谭庆昌　吉林大学

唐任仲　浙江大学

王连弟　华中科技大学出版社

吴鹿鸣　西南交通大学

杨玉虎　天津大学

赵永生　燕山大学

朱如鹏　南京航空航天大学

竺志超　浙江理工大学

# 前　言

　　材料力学实验教学是材料力学课程的一个重要环节。近年来,基于国际化的全英文教学在全国很多高校的一些专业进行了尝试,为配合全英文的材料力学实验课程的教学,浙江理工大学机械设计制造系力学课程组在原有《材料力学实验》讲义的基础上,根据教育部高等学校工科基础课程教学指导委员会公布的《材料力学课程教学基本要求(A 类)》(2019 版),结合教学的实际情况,编写了本实验教材。在本书编写过程中,编者结合前期在全英文班级和留学生教学试行中的反馈,对原有讲义重新进行了修订,对部分实验内容进行了调整,并将其整理为中英文对照版,供全英文班级和留学生使用。

　　本教材共分为两章,包含 9 个实验。第一章为材料的力学性能实验,包含的实验为:①金属材料的拉伸实验;②金属材料的扭转实验。第二章为电测法应力分析实验,包含的实验为:①电测法基础及应变仪使用;②矩形截面梁弯曲正应力测定实验(1/4 桥);③矩形截面梁弯曲正应力测定实验(半桥与全桥);④弯扭组合时的主应力主方向测定;⑤应变片粘贴实验;⑥等强度梁的测试实验;⑦梁的冲击载荷系数测定。

　　本教材所包含的实验内容,可以使学生得到以下三方面的训练:

　　(1) 学会测定金属材料的力学性能;

　　(2) 学会使用应变仪测定构件的应力和变形;

　　(3) 学会通过实验结果来验证基本理论。

　　本书包含了测定金属材料力学性能及基于电测法测试应力的多个实验,涵盖了高等学校本科机械类专业材料力学教学的内容,教师可根据实际情况对教学内容进行适当取舍。

　　本书可作为高等学校本科机械类各专业的材料力学实验课程教材,也可在各高校机械类专业国际化全英文教学中使用。

　　全书由浙江理工大学魏义敏主编,参加本书编写工作的还有乐忠萍、周迅、李剑敏等。具体编写分工为:魏义敏负责英文部分的编写和全书统稿;乐忠萍负责中文部分的编写;周迅提供了教材中的图片以及制作编辑表格;李剑敏对全书进行了审阅。

　　限于作者水平,加之时间紧迫,该教材难免有疏漏和不妥之处,恳请广大读者提出批评和指正,以便再版时修订。

编　者

2020 年 8 月

# 目　　录

# 绪　言

## 一、本课程的主要内容

材料力学实验是学习材料力学课程的一个重要环节。通过进行材料力学实验不仅能巩固和加深理解材料力学课程中的相关理论知识,学习各种实验仪器的使用方法,掌握常用材料的力学性能及构件的应力与变形的测试方法;同时还能够培养学生的动手能力,对数据结果的分析和处理能力,帮助学生树立起严肃认真的学习作风。

材料力学实验按性质可以分为以下三类。

### 1. 测定材料的力学性能

材料的力学性能是指在力的作用下,材料在变形、强度等方面表现出的一些特性,如弹性极限、屈服强度、弹性模量、疲劳强度、冲击韧度等。这些力学性能指标是构件强度、刚度和稳定性计算的依据。通过拉伸和扭转实验可以进一步使学生巩固有关材料力学性质的知识和掌握材料特性数值的基本测定方法。对材料力学性能的测定,应按国家标准(或行业标准)的规定进行。

### 2. 验证基本理论

将实际问题抽象为理想模型,再根据科学假设推导出一般性公式,这是材料力学中常采用的研究方法,例如杆件的弯曲理论就是以平面假设为基础的。但这些简化和假设是否正确,理论公式是否能在设计中应用,都需要通过实验来验证,如梁的弯曲正应力实验就属于这类实验。新建立的理论和公式,用实验来验证更是必不可少的。实验是验证、修正和发展理论的必要手段。学生通过这些实验可以巩固和加深理解课堂中所学的基本理论。

### 3. 测定构件的应力和变形

工程中很多实际构件的形状、受力情况、工作环境都十分复杂,如汽车底盘、水坝、水压机、飞机结构等。它们的应力及变形单纯靠理论计算是较难获得的,虽然有限元法和计算机的广泛应用,给计算法提供了有力的手段,但有时也不易得到满意的结果。实验应力分析的方法,是用实验的方法解决应力分析的问题。这类实验的方法很多,在我们的实验中,着重训练用电测法测量应变及应力。通过电测实验,让学生初步掌握电阻应变仪的使用及了解实测方法,提高解决实际问题的能力。

## 二、实验课的规则和要求

为了提高实验效果,保证实验课顺利进行,对参加实验的学生提出下列要求:

(1)实验课前必须按要求认真预习,按要求写好预习报告。通过预习,应明确本次实验的目的、原理及实验所需的仪器和实验方法。

(2)进入实验室要遵守实验室规则,不得大声喧哗,不得使用与本次实验无关的仪器设备。

(3)实验态度应严肃认真、科学严谨。实验开始时,每个实验小组成员应有明确分工,密切配合,相互协调。测量数据要真实有效。

(4)要爱护实验仪器及其他一切设备。材料力学实验所用的仪器是较为复杂和精密的,为了确保仪器设备的正常使用及其精密度,学生还必须认真阅读仪器操作规程,实验时要严格遵照规程进行操作,要注意安全,如发生故障应及时报告教师。

(5)实验完毕后,要认真检查实验记录,经实验指导教师审阅后,方可离开实验室。

## 三、实验报告注意事项

实验报告是实验资料的总结,应注意以下内容:

(1)在实验测量中,应注意测量单位。此外,还应注意仪器本身的精度。仪器的最小刻度值代表仪器的精度。多次测量同一物理量,每次所得数据并不完全相同,应以测量结果的算术平均值作为该物理量的测量值。

(2)报告中的最后结论,应是对实验中所观察到的主要现象和实验结果进行分析、总结和归纳而得出的。

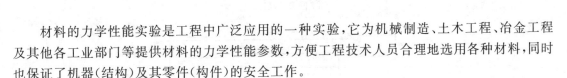

材料的力学性能实验是工程中广泛应用的一种实验,它为机械制造、土木工程、冶金工程及其他各工业部门等提供材料的力学性能参数,方便工程技术人员合理地选用各种材料,同时也保证了机器(结构)及其零件(构件)的安全工作。

## 实验一 金属材料的拉伸实验

### 一、实验目的

(1)测定低碳钢拉伸时的强度及塑性性能指标:下屈服强度 $\sigma_{eL}$、抗拉强度 $\sigma_m$、断后伸长率 $A$、断面收缩率 $Z$。

(2)测定灰铸铁拉伸时的强度性能指标:抗拉强度 $\sigma_m$。

(3)比较低碳钢与灰铸铁在拉伸时的力学性能和破坏形式。

### 二、实验设备和仪器

(1)电子万能试验机。

(2)游标卡尺。

### 三、实验试样

根据国家标准 GB/T 228.1—2010《金属材料 拉伸试验 第 1 部分:室温试验方法》,金属拉伸试样随着产品的品种、规格及试验目的的不同,其横截面形状可分为圆形、矩形、多边形、环形等,特殊情况下也可以为某些其他形状,其中最常用的是圆形截面试样和矩形截面试样。

如图 1-1 所示,圆形截面试样和矩形截面试样均由平行、过渡和夹持三部分组成。试样平行部分的长度记为 $L_c$,对于未经加工的试样,平行长度即为两夹头之间的距离。试样的原始标距 $L_0$ 与横截面面积 $S_0$ 之间满足 $L_0 = k \sqrt{S_0}$ 关系的试样称为比例试样,不满足的则为非比例试样。$k$ 值一般取 5.65,且原始标距应不小于 15 mm。当试样横截面面积过小时,$k$ 值可取 11.3,或者采用非比例试样。试样过渡部分以圆弧与平行部分光滑地连接,以保证试样断裂时的断口在平行部分。夹持部分稍大,其形状和尺寸根据试样大小、材料特性、试验目的及万能试验机的夹具结构进行设计。试样的形状、尺寸和加工的技术要求参见国家标准 GB/T 228.1—2010。

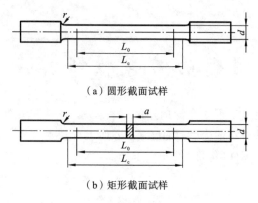

（a）圆形截面试样

（b）矩形截面试样

图 1-1　拉伸实验中的常见试样

## 四、实验原理

### 1. 低碳钢拉伸

低碳钢是典型的塑性材料,其拉伸过程大致分为四个阶段,即弹性阶段、屈服阶段、强化阶段、颈缩阶段。根据图 1-2 可以测定低碳钢拉伸弹性模量 $E$、下屈服载荷 $F_{eL}$,以及最大载荷 $F_m$。

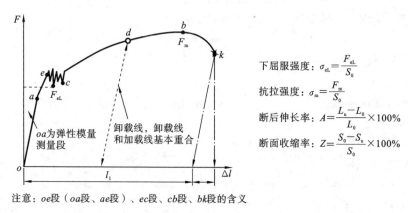

下屈服强度：$\sigma_{eL} = \dfrac{F_{eL}}{S_0}$

抗拉强度：$\sigma_m = \dfrac{F_m}{S_0}$

断后伸长率：$A = \dfrac{L_u - L_0}{L_0} \times 100\%$

断面收缩率：$Z = \dfrac{S_0 - S_u}{S_0} \times 100\%$

$oa$ 为弹性模量测量段

卸载线,卸载线和加载线基本重合

注意：$oe$ 段（$oa$ 段、$ae$ 段）、$ec$ 段、$cb$ 段、$bk$ 段的含义

图 1-2　低碳钢的拉伸曲线

图 1-2 中,$L_0$ 为试样的原始标距,$L_u$ 为试样的断后标距,$S_0$ 为试样的原始横截面面积,$S_u$ 为试样断裂后的最小横截面面积。试样的塑性变形集中产生在颈缩处,并向两边逐渐减小。因此,断口的位置不同,标距 $L_0$ 部分的塑性伸长量也不同。若断口在试样的中部,则发生严重塑性变形的颈缩段全部在标距长度内,标距长度就有较大的塑性伸长量;若断口离标距端很近,则发生严重塑性变形的颈缩段只有一部分在标距长度内,另一部分在标距长度外,在这种

情况下,标距长度的塑性伸长量就较小。因此,断口的位置对所测得的断后伸长率有影响。为了避免这种影响,国家标准对断后标距 $L_u$ 的测定做了如下规定:测量时,两段在断口处应紧密对接,尽量使两段的轴线在一条直线上。若断口处有缝隙,则此缝隙应计入 $L_u$ 内;如果断口在标距长度以外,或者虽在标距长度之内,但离标距端的距离小于 $2d$,则实验无效。

弹性模量 $E$ 的测定:材料在线弹性范围内,其应力 $\sigma$、应变 $\varepsilon$ 成正比关系,即

$$\sigma_p = E \cdot \varepsilon_p \rightarrow E = \frac{\Delta F_p \cdot L_0}{S_0 \cdot \Delta L}$$

### 2. 灰铸铁拉伸

灰铸铁拉伸过程比较简单,可近似认为经弹性阶段直接过渡到断裂阶段,没有屈服现象和颈缩现象,具体如图 1-3 所示。断后伸长率和断面收缩率极小,灰铸铁的抗拉强度远低于它的抗压强度,是典型的脆性材料。

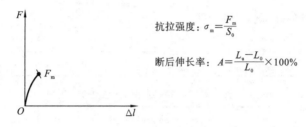

抗拉强度:$\sigma_m = \frac{F_m}{S_0}$

断后伸长率:$A = \frac{L_u - L_0}{L_0} \times 100\%$

图 1-3 灰铸铁拉伸曲线

## 五、实验步骤

(1) 测量试样的尺寸。选取试样,每组一根低碳钢、一根灰铸铁,可听音判别材质,并擦拭干净;用游标卡尺测量试样的直径和标距,并记录数据。

(2) 拉伸试样。按操作规程使用电子万能试验机拉伸试样,观察材料在拉伸时的现象,直至试样被拉断,记录载荷数据。

(3) 测量试样断后尺寸。取下试样,将断口吻合压紧,用游标卡尺量取断后直径和标距。

(4) 利用电子万能试验机开展实验。具体操作规程如下:

① 开机,打开试验机主开关,然后打开计算机,进入操作界面。

② 根据试样的形状、尺寸更换合适的夹具。

③ 通过门柱上的控制盒,移动上横梁到合适位置。先装好上夹头,然后移动横梁,装下夹头。

④ 在计算机软件中新建试样,输入试样信息。试样编号:低碳钢 XY1,灰铸铁 XY2(X 为批次,Y 为组次)。

⑤ 选择合适的速度。将位移和时间清零,点击界面上的"力清零",然后开始拉伸。对于低碳钢,材料屈服前可采用 $2\sim5$ mm/min 的速度;材料进入强化阶段后,用速度滑块均匀加速到 $10\sim20$ mm/min。对于其他材料,按照相应标准所规定的速度进行实验。

⑥ 试样断裂后,保存实验数据和结果,生成报告;然后在计算机软件中进行实验分析→实验报告→获取拉伸图形和载荷数据。

⑦ 取下试样,进行下一个试样的实验。

## 六、实验数据记录与计算

将低碳钢拉伸和灰铸铁拉伸的实验数据分别记录于表 1-1 和表 1-2。

表 1-1　低碳钢拉伸的实验数据

| 实验数据 | | | 第一次测量 | 第二次测量 | 平均值 |
|---|---|---|---|---|---|
| 拉伸前 | 截面一直径 | $d_{01}$/mm | | | |
| | 截面二直径 | $d_{02}$/mm | | | |
| | 截面三直径 | $d_{03}$/mm | | | |
| | 原始标距 | $L_0$/mm | | | |
| 拉伸后 | 最小截面直径 | $d_1$/mm | | | |
| | 断后标距 | $L_u$/mm | | | |
| | 下屈服载荷/kN | $F_{eL}$ | | | |
| | 最大载荷/kN | $F_m$ | | | |

表 1-2　灰铸铁拉伸的实验数据

| 实验数据 | | | 第一次测量 | 第二次测量 | 平均值 |
|---|---|---|---|---|---|
| 拉伸前 | 截面一直径 | $d_{01}$/mm | | | |
| | 截面二直径 | $d_{02}$/mm | | | |
| | 截面三直径 | $d_{03}$/mm | | | |
| | 原始标距 | $L_0$/mm | | | |
| 拉伸后 | 最大载荷/kN | $F_m$ | | | |

### 1. 计算精确度

(1) 强度性能指标(下屈服强度 $\sigma_{eL}$ 和抗拉强度 $\sigma_m$)的计算精度要求为 0.5 MPa,即舍去所有小于 0.25 MPa 的数值,大于或等于 0.25 MPa 而小于 0.75 MPa 的数值化为 0.5 MPa,大于或等于 0.75 MPa 的数值则进为 1 MPa。

(2) 塑性性能指标(断后伸长率 $A$ 和断面收缩率 $Z$)的计算精度要求为 0.5%,即舍去所有小于 0.25% 的数值,大于或等于 0.25% 而小于 0.75% 的数值化为 0.5%,大于或等于 0.75% 的数值则进为 1%。

### 2. 问题讨论和分析

(1) 低碳钢拉伸试样断口特点:低碳钢拉伸试样断口呈杯锥状,有颈缩现象。断口中部区域粗糙,呈脆性断裂(被拉断),断口外围光滑,是塑性变形区域,有 45° 的剪切唇(被剪断)。

(2) 灰铸铁拉伸试样断口特点:灰铸铁拉伸试样断口呈凹凸不平整的颗粒状,整个断面大约与试样轴线垂直,没有颈缩现象,是典型的脆性断裂(被拉断)。

## 七、注意事项

实验时严禁开"快速"挡加载。加载时速度要均匀缓慢,防止冲击。实验中如发生故障应立即停机,报告老师。

## 八、思考题

(1)画出低碳钢和灰铸铁拉伸的载荷变形曲线图,并标注关键点。

(2)分析低碳钢和灰铸铁在常温静载拉伸时的力学性能和破坏形式的异同。

(3)测定材料的力学性能有何实用价值?

(4)你认为使实验结果产生误差的因素有哪些? 应如何避免或减小其影响?

# 实验二　金属材料的扭转实验

## 一、实验目的

(1) 测定金属材料扭转时的强度性能指标：低碳钢扭转的下屈服强度 $\tau_{eL}$ 和抗扭强度 $\tau_m$；灰铸铁的抗扭强度 $\tau_m$。

(2) 绘制低碳钢和灰铸铁的断口图，比较低碳钢和灰铸铁的扭转破坏形式。

## 二、实验设备和仪器

(1) 扭转试验机。

(2) 游标卡尺。

## 三、实验试样

按照国家标准 GB/T 10128—2007《金属材料　室温扭转试验方法》，金属扭转试样随着产品的品种、规格以及试验目的的不同可分为圆形截面试样和管形截面试样两种，其中，最常用的是圆形截面试样，如图 1-4 所示。通常，圆形截面试样的推荐直径为 $d=10$ mm，标距 $L_0$ 为 50 mm 或者 100 mm，平行部分的长度为 $L_0+2d$。若采用其他直径的试样，其平行部分的长度也应为标距加上两倍直径。试样头部的形状和尺寸应适合扭转试验机的夹头夹持。

图 1-4　圆形截面试样

由于在扭转实验中，试样表面的切应力最大，试样表面的缺陷将敏感地影响实验结果，因此，对扭转试样的表面粗糙度的要求比拉伸试样的高。扭转试样的加工技术要求参照国家标准。

## 四、实验原理

### 1. 测定低碳钢扭转时的性能指标

试样在外力偶矩的作用下，其上任意一点都处于纯剪切应力状态。随着外力偶矩的增大，试样应力也增大，试样由弹性状态到屈服状态，再到强化状态，最后被扭断，具体如图 1-5 所示。低碳钢扭转的下屈服强度为

$$\tau_{eL} = \frac{3}{4} \frac{T_{eL}}{W}$$

式中:$W = \pi d^3/16$ 为试样在标距内横截面的抗扭截面系数。

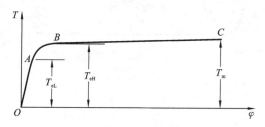

图 1-5  低碳钢的扭转曲线

可以求得,低碳钢的抗扭强度为

$$\tau_m = \frac{3}{4} \frac{T_m}{W}.$$

由图 1-6 可知,圆柱形试样横截面承受的扭矩为

$$T_{eL} = \int_0^{d/2} \tau_{eL}\rho 2\pi\rho d\rho = 2\pi\tau_{eL} \int_0^{d/2} \rho^2 d\rho = \frac{\pi d^3}{12} \tau_{eL} = \frac{4}{3} W \tau_{eL}$$

由上式可以求得 $\tau_{eL}$。

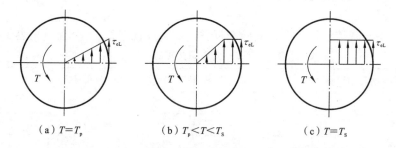

（a）$T = T_p$　　　（b）$T_p < T < T_s$　　　（c）$T = T_s$

图 1-6  低碳钢圆柱形试样扭转时横截面上的切应力分布

从图 1-5 可知,当外力偶矩超过 $T_{eL}$ 后,随着扭转角 $\varphi$ 的增大,外力偶矩 $T$ 增大幅度很小,$BC$ 近似于一条直线。因此,可认为横截面上的切应力分布如图 1-6 所示,只是切应力值比 $\tau_{eL}$ 大。根据测定的试样在断裂时的外力偶矩 $T_m$,可求得抗扭强度为

$$\tau_m = \frac{3}{4} \frac{T_m}{W}$$

## 2. 测定灰铸铁扭转时的强度性能指标

对于灰铸铁试样,只需测出其承受的最大外力偶矩 $T_m$,即可求得其抗扭强度为

$$\tau_m = \frac{T_m}{W}$$

由扭转破坏的试样可以看出:低碳钢试样的断口与轴线垂直,表明破坏是由切应力引起的;而灰铸铁试样的断口则沿螺旋线方向与轴线约成 45° 角,表明破坏是由拉应力引起的。

试件受扭,材料处于纯剪应力状态,具体如图 1-7 所示。与试样轴线成 45° 角的螺旋面分别受到主应力 $\sigma_1$ 和 $\sigma_3$ 的作用。低碳钢的抗拉能力大于抗剪能力,故在横截面处剪断,而灰铸铁的抗拉能力比抗剪能力弱,故沿着与轴线约成 45° 角方向被拉断。

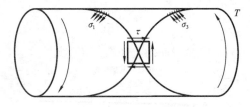

图 1-7 纯剪应力状态

## 五、实验步骤

（1）测量试样尺寸。在标距的两端和中间三个位置上，沿互相垂直的方向，测量试样的直径，并计算每个截面的平均直径。选用三处中的最小直径来计算原始横截面面积 $S_0$、标距 $L_0$。

（2）装夹试样。先装移动端夹头，扭矩清零，然后安装固定端。

（3）操作计算机。选择方案，输入试样信息。

（4）加载。按扭转试验机的操作规程进行。

（5）试样扭断自动停机。保存数据，记录图形和扭矩。

## 六、实验数据记录与计算

记录低碳钢和灰铸铁扭转时的实验数据并测定强度性能指标，分别填入表 1-3 和表 1-4。

表 1-3　低碳钢扭转的数据

| 低碳钢实验数据 | | 第一次测量 | 第二次测量 | 平均值 |
|---|---|---|---|---|
| 截面一直径 | $d_{01}/mm$ | | | |
| 截面二直径 | $d_{02}/mm$ | | | |
| 截面三直径 | $d_{03}/mm$ | | | |
| 屈服扭矩/$(N \cdot m)$ | $T_{eL}$ | | | |
| 最大扭矩/$(N \cdot m)$ | $T_m$ | | | |

表 1-4　灰铸铁扭转的数据

| 灰铸铁实验数据 | | 第一次测量 | 第二次测量 | 平均值 |
|---|---|---|---|---|
| 截面一直径 | $d_{01}/mm$ | | | |
| 截面二直径 | $d_{02}/mm$ | | | |
| 截面三直径 | $d_{03}/mm$ | | | |
| 最大扭矩/$(N \cdot m)$ | $T_m$ | | | |

## 七、思考题

（1）画出低碳钢和灰铸铁的扭转曲线图，并标出关键点。

（2）画出低碳钢和灰铸铁的断口示意图。

（3）比较低碳钢与灰铸铁试样的扭转破坏断口，试分析它们的破坏原因。

# 第二章

## 电测法应力分析实验

电测法是应力分析中应用最为广泛和有效的方法之一,可应用于机械、土木、水利、航空航天等技术领域,是验证材料力学基本理论、检验工程质量和开展科学研究的有效手段之一。

## 实验三　电测法基础及应变仪使用

**电阻应变测量方法**是将应变转换成电信号进行测量的方法,简称**电测法**。电测法的基本原理是:将电阻应变片(简称**应变片**)粘贴在被测构件的表面,当构件发生变形时,应变片随着构件一起变形,应变片的电阻值将发生相应的变化,利用电阻应变测量仪器(简称**电阻应变仪**)测量出应变片电阻值的变化,并将其换算成应变值,或输出与应变成正比的模拟电信号(电压或电流),用记录仪记录下来,实现对应变的测量。

应变片质量轻、体积小且可在高(低)温、高压等特殊环境下使用,在利用电测法进行测量时输出量为电信号,便于实现自动化和数字化,这使得电测法具有很高的灵敏度,并能实现远距离测量及无线测量。

### 一、电阻应变片结构

电阻应变片的构造很简单,把一根很细的具有高电阻率的金属丝在制片机上按图 2-1 所示排绕后,用胶水粘贴在两个薄片之间,再焊上较粗的引出线,就成了早期常用的丝绕式应变片。应变片一般由敏感栅(即金属丝)、粘接剂、基底、引出线和覆盖层五部分组成。将应变片粘贴在被测构件的表面后,金属丝随构件一起变形时,其电阻值也随之变化。

常用的应变片有丝绕式应变片(见图 2-1)、短接线式应变片和箔式应变片(见图 2-2)等。它们均属于单轴式应变片,即一个基底上只有一个敏感栅,用于测量沿栅轴方向的应变。如图

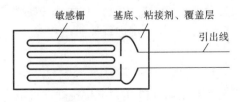

图 2-1　应变片的构造

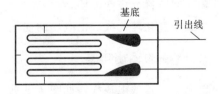

图 2-2　箔式应变片

2-3 所示,有的应变片在同一基底上按一定角度布置了几个敏感栅,可测量同一点沿几个敏感栅栅轴方向的应变,因而称为**多轴应变片**,俗称**应变花**。应变花主要用于测量平面应力状态下某一点的主应变和主方向。

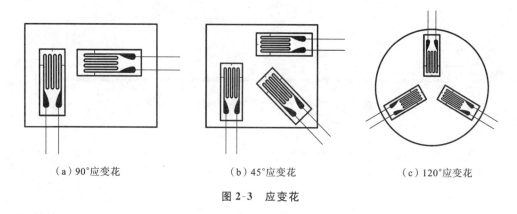

（a）90°应变花        （b）45°应变花        （c）120°应变花

图 2-3    应变花

## 二、电阻应变片的灵敏系数

在用应变片进行应变测量时,需要给应变片中的金属丝上电以进行测试。为了防止电流过大,产生发热和熔断等现象,要求金属丝有一定的长度,以使其具有较大的初始电阻值。但在测量构件的应变时,又要求尽可能缩短金属丝的长度,以测得"某一点"的真实应变。因此,应变片中的金属丝一般做成图 2-1 所示的栅状,称为敏感栅。粘贴在构件上的应变片,其金属丝的电阻值随着构件的变形而发生变化的现象称为电阻应变现象。当把应变片安装在处于单向应力状态的试件表面并使敏感栅的栅轴方向与应力方向一致时,在一定的变形范围内,应变片电阻值的变化率 $\Delta R/R_0$ 与敏感栅栅轴方向的应变 $\varepsilon$ 成正比,即

$$\frac{\Delta R}{R_0} = K\varepsilon$$

式中:$R_0$ 为应变片的原始电阻值;$\Delta R$ 为应变片电阻值的改变量;$K$ 称为应变片的灵敏系数。

应变片的灵敏系数一般由制造厂家通过实验测定,这一步骤称为应变片的标定。在实际应用时,可根据需要选用不同灵敏系数的应变片。

## 三、电阻应变片的测量电路

在使用应变片测量应变时,必须用适当的办法测量其电阻值的微小变化。为此,一般把应变片接入某一电路,让其电阻值的变化使电路改变,使该电路输出一个能反映应变片电阻值变化的信号,然后,对该电信号进行相应的处理即可获得应变值。通常,电测法所使用的电阻应变片输入回路称为应变电桥,它是以应变片作为其部分或全部桥臂的四臂电桥,能把应变片电阻值的微小变化转化成输出电压的变化。本书以直流电压电桥为例进行说明。

### 1. 电桥的输出电压

电阻应变片的电桥线路如图 2-4 所示,它以应变片或电阻元件作为电桥桥臂。可取 $R_1$ 为应变片、$R_1$ 和 $R_2$ 为应变片或 $R_1 \sim R_4$ 均为应变片等几种形式,$A$、$C$ 和 $B$、$D$ 分别为电桥的输

入端和输出端。

根据电工学原理,可导出当输入端加有电压 $U_1$ 时,电桥的输出电压为

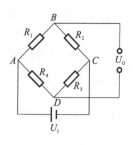

图 2-4　电桥线路

$$U_O = \frac{R_1 R_3 - R_2 R_4}{(R_1 + R_2)(R_3 + R_4)} U_1$$

当 $U_O = 0$ 时,电桥处于平衡状态。因此,电桥的平衡条件为 $R_1 R_3 = R_2 R_4$。当处于平衡的电桥中各桥臂的电阻值的变化分别为 $\Delta R_1$、$\Delta R_2$、$\Delta R_3$ 和 $\Delta R_4$ 时,可近似地求得电桥的输出电压为

$$U_O \approx \frac{U_1}{4}\left(\frac{\Delta R_1}{R_1} - \frac{\Delta R_2}{R_2} + \frac{\Delta R_3}{R_3} - \frac{\Delta R_4}{R_4}\right)$$

由此可见,应变电桥有一个重要的性质:应变电桥的输出电压与相邻两桥臂的电阻变化率之差、相对两桥臂电阻变化率之和成正比。对于平衡电桥,如果相邻两桥臂的电阻变化率大小相等、符号相同,或相对两桥臂的电阻变化率大小相等、符号相反,则电桥的平衡状态将不会改变,即保持 $U_O = 0$。如果电桥的四个桥臂均接入相同的应变片,则有

$$U_O = \frac{K U_1}{4}(\varepsilon_1 - \varepsilon_2 + \varepsilon_3 - \varepsilon_4)$$

式中:$\varepsilon_1 \sim \varepsilon_4$ 分别为接入电桥四个桥臂的应变片的应变值。

### 2. 温度效应的补偿

贴有应变片的构件总是处在某一温度场中,若敏感栅材料的线膨胀系数与构件材料的线膨胀系数不相等,则当温度发生变化时,由于敏感栅与构件的伸长(或缩短)量不相等,敏感栅会受到附加的拉伸(或压缩)作用,从而引起敏感栅电阻值的变化,这种现象称为温度效应。敏感栅电阻值随温度的变化率可近似地看作与温度成正比。温度的变化对电桥的输出电压影响很大,严重时,每升温 1 ℃,电阻应变片中可产生几十微应变($\mu\varepsilon$,$\mu\varepsilon$ 指应变的量级,力学中的惯用语),因此需要采取相关措施来消除温度效应。消除温度效应的措施,称为温度补偿。

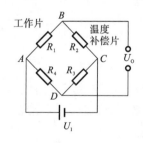

图 2-5　半桥单臂接法

根据电桥的性质,温度补偿并不困难。只要用一个应变片作为温度补偿片,将它粘贴在一块与被测构件材料相同但不受力的试件上。将此试件和被测构件放在一起,使它们处于同一温度场中。粘贴在被测构件上的应变片称为工作片。在连接电桥时,使工作片与温度补偿片处于相邻的桥臂,如图 2-5 所示。因为工作片和温度补偿片的温度始终相同,所以它们因温度变化而引起的电阻值的变化也相同,又因为它们处于电桥相邻的两臂,从而消除了温度效应的影响。

工作片和温度补偿片的电阻值、灵敏系数及电阻温度系数应相同,分别粘贴在被测构件和同材质不受力的试件上,以保证它们因温度变化而引起的应变片电阻值的变化相同。

### 3. 应变片的布置和在电桥中的接法

应变片反映的是构件表面某点的拉应变或压应变。在有些情况下,该应变可能与多种内力(如轴力、弯矩等)有关。一般情况下,只需测量出与某种内力对应的应变,而要把与其他内力对应的应变从总应变中排除掉。显然,应变片本身不会分辨应变成分,但是只要合理地选择粘贴应变片的位置和方向,并把应变片合理地接入电桥,就能利用电桥的性质,从比较复杂的

组合应变中测量出指定"某一点"的应变。

应变片在电桥中的接法常有以下三种形式：

（1）**半桥单臂接法**   又名 1/4 桥接法。如图 2-5 所示,将一个工作片和一个温度补偿片分别接入两个相邻桥臂,另两个桥臂接固定电阻。设工作片的应变为 $\varepsilon_1$,则电桥的输出电压为

$$U_O = \frac{KU_1}{4}\varepsilon_1$$

应变仪读数 $\varepsilon = \varepsilon_1$。

（2）**半桥双臂接法**   如图 2-6 所示,将两个工作片接入电桥的两个相邻桥臂,另两个桥臂接固定电阻,两个工作片同时互为温度补偿片。设工作片的应变分别为 $\varepsilon_1$ 和 $\varepsilon_2$,$\varepsilon_1 = -\varepsilon_2$,则电桥的输出电压为

$$U_O = \frac{KU_1}{4}(\varepsilon_1 - \varepsilon_2) = \frac{KU_1}{2}\varepsilon_1$$

应变仪读数 $\varepsilon = 2\varepsilon_1$。

（3）**全桥接法**   如图 2-7 所示,电桥的四个桥臂全部接入工作片,设工作片的应变分别为 $\varepsilon_1$、$\varepsilon_2$、$\varepsilon_3$ 和 $\varepsilon_4$,且 $\varepsilon_1 = -\varepsilon_2 = \varepsilon_3 = -\varepsilon_4$,则电桥的输出电压为

$$U_O = \frac{KU_1}{4}(\varepsilon_1 - \varepsilon_2 + \varepsilon_3 - \varepsilon_4) = KU_1\varepsilon_1$$

应变仪读数 $\varepsilon = 4\varepsilon_1$。

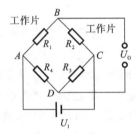 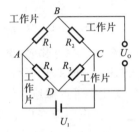

图 2-6   半桥双臂接法            图 2-7   全桥接法

## 四、静态电阻应变仪

静态电阻应变仪是专门测量不随时间变化或变化极缓慢的电阻应变的仪器,其功能是将应变电桥的输出电压放大,在显示部分以刻度或数字形式显示应变的数值,或者向记录仪输出模拟应变变化的电信号。

### 1. 静态电阻应变仪的工作原理

静态电阻应变仪的种类很多,本书以 CML-1H 型静态电阻应变仪为例来介绍其工作原理。

CML-1H 型静态电阻应变仪的工作原理框图如图 2-8 所示,它内置单片机,可完成采集、处理、显示、通信等功能,同时也可进行 1/4 桥、半桥或全桥测量。在测量应变时,把粘贴在构件上的应变片接入电桥,将电桥预调平衡,当构件受力发生变形时,应变片电阻值随之变化,从而电桥的平衡被破坏,产生输出电压,由显示表显示出应变的数值。该应变仪可同时测量 16个点的应变,还可以通过接口与计算机连接,由计算机对测量数据进行处理。

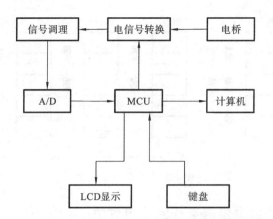

图 2-8　CML-1H 型静态电阻应变仪的工作原理框图

## 2. CML-1H 型静态电阻应变仪电桥的接线方式

电阻应变仪电桥的接线方式如图 2-9 至图 2-12 所示。

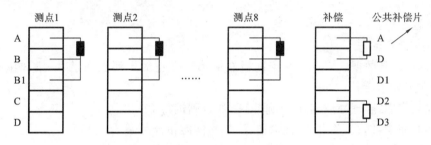

图 2-9　1/4 桥接线方式

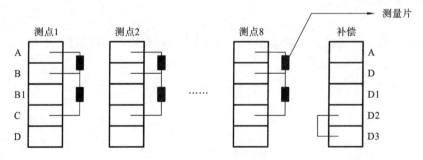

图 2-10　半桥接线方式

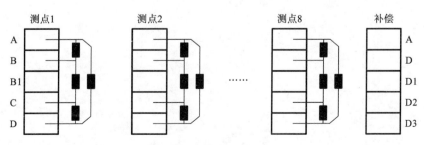

图 2-11　全桥接线方式

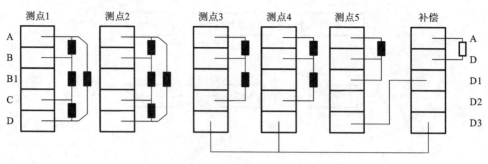

图 2-12    混合接线方式

### 3. CML-1H 型静态电阻应变仪的使用方法

（1）开机。接通 220 V 电源并打开应变仪的开关，此时 9 组数码管发亮，由 5 到 0 递减显示，完成仪器自检后机号显示位闪烁，直接点击"确定"键，预热 30 min。

（2）预热，同时接线。按实验要求选择所需桥路，对照接线图，将应变片连接到电阻应变仪的各路端子上。

（3）设置参数。

（4）应变清零。

（5）加载，记录数据。应变仪数字面板左侧第 1、2 位显示测点通道号，第 3 位显示正负号，第 4～8 位显示应变值。

（6）加载完成，整理好数据，给老师审阅，通过则进行下一步。

（7）实验完毕，卸载，关电源，拆除连线，整理仪器和台面，一切复原。

## 五、应变仪接线实验

以等截面梁的弯曲正应力实验装置为例，进行半桥单臂（1/4 桥）接法、半桥双臂接法和全桥接法练习。

## 实验四 矩形截面梁弯曲正应力测定实验(1/4桥)

### 一、实验目的

(1) 熟悉电测法的基本原理和静态电阻应变仪的使用方法。
(2) 测量矩形截面梁在纯弯曲和横力弯曲时横截面上的正应力分布。

### 二、实验设备和仪器

(1) WYS-1型材料力学实验台。
(2) 静态电阻应变仪。

### 三、实验原理

实验装置如图2-13所示,矩形截面梁采用低碳钢制成。在梁发生纯弯曲和横力弯曲变形段的侧面上,分别在与轴线平行的不同高度的线段上各粘贴8个应变片作为工作片,另外在与梁同材料的钢片上粘贴2个应变片作为温度补偿片。

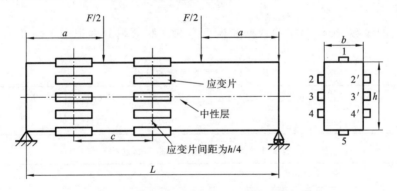

**图2-13 等截面梁的弯曲正应力实验装置示意图**

将16个工作片和温度补偿片以1/4桥形式分别接入电阻应变仪面板上的通道中。当梁在载荷作用下发生弯曲变形时,工作片的电阻值将随着梁的变形而发生变化,通过电阻应变仪可以分别测量出各对应位置的应变值 $\varepsilon_{实}$,应变沿梁高度的分布如图2-14所示。

实验中采用等量逐级加载法。根据胡克定律,计算各点的实验正应力:

$$\sigma_{实}=E\varepsilon_{实}$$

式中:$E$ 为梁材料的弹性模量。

梁在弯曲变形时,横截面上的理论正应力为

$$\sigma_{理}=\frac{My}{I_z}$$

式中:$I_z=bh^3/12$,为梁的横截面对中性轴的惯性矩;$y$ 为中性轴到各应力点的距离。对于纯

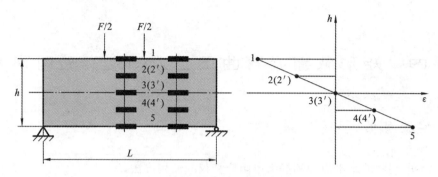

**图 2-14　应变沿梁高度的分布**

变曲变形,$M_{纯}=Fa/2$;对于横力弯曲变形,$M_{横}=F(a-c/2)/2$。

## 四、实验步骤

(1) 接线。按 1/4 桥接法接线。操作步骤参见应变仪操作规程。

(2) 加载。摇动手轮匀速缓慢加载,采用等量逐级加载(取 $F_1=1$ kN,增量 0.5 kN,$F_{\max}=3$ kN)的方法,每增加一级载荷,就记录各点的应变值(注意正负号)。

(3) 实验完毕后,缓慢卸载,将数据交给老师检查。然后关电源,拆除连线,整理仪器使之归位。

## 五、实验数据记录与计算

将截面梁的参数、纯弯曲实验数据和横力弯曲实验数据分别记录于表 2-1、表 2-2 和表 2-3 中。

**表 2-1　矩形截面梁的参数**

| $a$/mm | $b$/mm | $c$/mm | $L$/mm | $h$/mm | $E$/GPa | $K$ | $R_0$ |
|--------|--------|--------|--------|--------|---------|-----|-------|
|        |        |        |        |        |         |     |       |

**表 2-2　矩形截面梁纯弯曲实验数据**

| 载荷/kN | 测点应变($\mu\varepsilon$) | | | | | | | |
|---------|-------------|-------------|---------------|---------------|----------------|----------------|---------------|----------------|
| | $y_1=0$ | $y_2=0$ | $y_3=h/4$ | $y_4=h/4$ | $y_5=-h/4$ | $y_6=-h/4$ | $y_7=h/2$ | $y_8=-h/2$ |
| 1 | | | | | | | | |
| 1.5 | | | | | | | | |
| 2 | | | | | | | | |
| 2.5 | | | | | | | | |
| 3 | | | | | | | | |
| $\overline{\varepsilon_i}=$ | $\overline{\varepsilon_1}=$ | $\overline{\varepsilon_2}=$ | $\overline{\varepsilon_3}=$ | $\overline{\varepsilon_4}=$ | $\overline{\varepsilon_5}=$ | $\overline{\varepsilon_6}=$ | $\overline{\varepsilon_7}=$ | $\overline{\varepsilon_8}=$ |
| $\sigma_{实}=E\overline{\varepsilon}$ | | | | | | | | |
| $\sigma_{理}=\dfrac{\overline{M}y}{I_z}$ | | | | | | | | |

表 2-3  矩形截面梁横力弯曲实验数据

| 载荷/kN | 测点应变($\mu\varepsilon$) | | | | | | | |
|---|---|---|---|---|---|---|---|---|
| | $y_1=0$ | $y_2=0$ | $y_3=h/4$ | $y_4=h/4$ | $y_5=-h/4$ | $y_6=-h/4$ | $y_7=h/2$ | $y_8=-h/2$ |
| 1 | | | | | | | | |
| 1.5 | | | | | | | | |
| 2 | | | | | | | | |
| 2.5 | | | | | | | | |
| 3 | | | | | | | | |
| $\overline{\varepsilon_i}=$ | $\overline{\varepsilon_1}=$ | $\overline{\varepsilon_2}=$ | $\overline{\varepsilon_3}=$ | $\overline{\varepsilon_4}=$ | $\overline{\varepsilon_5}=$ | $\overline{\varepsilon_6}=$ | $\overline{\varepsilon_7}=$ | $\overline{\varepsilon_8}=$ |
| $\sigma_{实}=E\overline{\varepsilon}$ | | | | | | | | |
| $\sigma_{理}=\dfrac{\overline{M}y}{I_z}$ | | | | | | | | |

## 六、注意事项

（1）检查矩形梁的位置，尽量使压头的中心线通过梁的纵向轴对称平面，以保证矩形梁的中部为纯弯曲变形。

（2）应变仪平衡操作时应确认梁上未加载荷，测力显示读数为"0"（如不为零则记录初始值）。

（3）加载时，手轮应平稳转动，不宜过快，待力值基本稳定在指定载荷值时，停止转动手轮并测量应变。

（4）加载的最大载荷为 3000 N，严禁超载，以免损坏力传感器。

（5）测量时，应保证接线不松动，在一个加载测量循环中不要移动和接触应变片的导线，以保证应变值测量数据的稳定可靠。

## 七、数据处理

（1）应变增量计算。根据记录的各点应变读数值，算出各点的 4 个应变增量。根据胡克定律，理论上同一测点测得的 4 个应变增量应相同，若同一测点的应变增量比较离散，要查找原因。

（2）各点实测应力计算。计算各点应变增量的平均值。对于在梁的前后侧面上都有应变片的测点，应把前后面的应变增量平均值进行再平均，得到该点在载荷增量为 0.5 kN 时的平均应变增量。由胡克定律 $\sigma=E\varepsilon$ 计算该点的应力实测值。由于电阻应变仪的最小应变读数为"1"，表示 1 $\mu\varepsilon$，即应变值为"$1\times10^{-6}$"。因此在计算实测应变增量时，最小数值也是 1 $\mu\varepsilon$。

（3）误差计算及分析。中性层（即 $y=0$ 处）的应力误差是绝对误差，其他点误差为相对误差。

## 八、思考题

(1) 为什么要把温度补偿片贴在与构件材料相同的材料上？

(2) 弯曲正应力的大小是否会受材料弹性模量的影响？为什么？

# 实验五 矩形截面梁弯曲正应力测定实验(半桥与全桥)

## 一、实验目的

(1)熟悉电测法的基本原理,学习半桥与全桥接线方法,进一步掌握测量多点静态应变的方法。

(2)用半桥和全桥方法测定等截面梁在纯弯曲和横力弯曲时横截面上的正应力,并与1/4桥的进行比较。

## 二、实验设备和仪器

(1)WYS-1型材料力学实验台。

(2)静态电阻应变仪。

(3)矩形截面单体梁。

## 三、实验原理

如图 2-15 所示,载荷 $F$ 通过加力梁均分成两个大小为 $F/2$ 的力作用在矩形钢梁上。梁的中部呈纯弯曲变形,弯矩为 $M=\dfrac{1}{2}Fa$。

在梁中部(纯弯曲)及加力梁外部(横力弯曲)的上、下表面及前、后两侧面以梁的中性层为基准,每隔 $h/4$ 处贴一枚平行于梁轴线的电阻应变片,合计 16 枚。

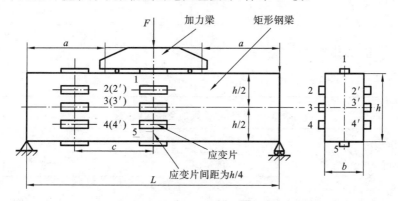

图 2-15 矩形截面梁的布片示意图

各枚应变片敏感栅的中心即为实验的测量点。根据各测量点的应变测量值,由胡克定律($\sigma=E\varepsilon$)计算应力实测值,得到横截面上正应力沿梁高的分布规律,并与弯曲正应力公式($\sigma=\dfrac{My}{I_z}$)计算的应力理论值进行比较。式中,$I_z=bh^3/12$ 为梁的横截面对中性轴的惯性矩,$y$ 为中性轴到各应力点的距离。对于纯弯曲变形,$M_纯=Fa/2$,对于横力弯曲变形,$M_横=F(a-c/2)/2$。

## 四、实验步骤

(1)接线。按半桥接法接线。操作步骤详见应变仪操作规程。

(2)加载。摇动手轮匀速缓慢加载,采用等量逐级加载(取 $F_1 = 1$ kN,增量 0.5 kN,$F_{max}$ = 3 kN)的方法,每增加一级载荷,就记录各点的应变值(注意正负号)。

(3)实验完毕后,缓慢卸载,将实验数据交给实验指导教师审阅,通过后再关电源,拆除连线。

(4)按全桥接法接线,重复以上步骤。

(5)实验完毕,缓慢卸载,关闭电源,拆除连线,整理好实验台。

## 五、实验数据记录与计算

将梁的尺寸、材料弹性模量 $E$、应变片灵敏系数 $K$ 和应变片电阻值 $R_0$ 记录于表 2-4。矩形截面梁的半桥实验数据和全桥实验数据记录于表 2-5 和表 2-6。

表 2-4 矩形截面梁参数

| $a$/mm | $b$/mm | $c$/mm | $L$/mm | $h$/mm | $E$/GPa | $K$ | $R_0$ |
|--------|--------|--------|--------|--------|---------|-----|-------|
|        |        |        |        |        |         |     |       |

表 2-5 矩形截面梁的半桥实验数据

| 载荷/kN | 测点应变($\mu\varepsilon$) | | | | | |
|---------|------|------|------|------|------|------|
|         | 纯弯曲 | | | 横力弯曲 | | |
|         | $y_1 =$ | $y_2 =$ | $y_3 =$ | $y_4 =$ | $y_5 =$ | $y_6 =$ |
| 1       |      |      |      |      |      |      |
| 1.5     |      |      |      |      |      |      |
| 2       |      |      |      |      |      |      |
| 2.5     |      |      |      |      |      |      |
| 3       |      |      |      |      |      |      |

表 2-6 矩形截面梁的全桥实验数据

| 载荷/kN | 测点应变($\mu\varepsilon$) | |
|---------|------|------|
|         | 纯弯曲 | 横力弯曲 |
|         | $y_1 =$ | $y_2 =$ |
| 1       |      |      |
| 1.5     |      |      |
| 2       |      |      |
| 2.5     |      |      |
| 3       |      |      |

## 六、注意事项

（1）进行应变仪平衡操作时应确认梁上未加载荷，测力显示读数为"0"（如不为零则记录初始值）。

（2）加载时，应检查加力梁和矩形钢梁的位置，尽量使力中心线通过梁的纵向轴对称平面，以保证矩形钢梁的中部为纯弯曲变形。

（3）加载时，手轮应平稳转动，不宜过快，待力值基本稳定在指定载荷值时，停止转动手轮并测量应变。

（4）加载的最大载荷为 3000 N，严禁超载，以免损坏力传感器。

（5）测量时，应保证接线不松动，在一个加载测量循环过程中请不要移动和接触应变片的导线，以保证测量数据的稳定可靠。

## 七、思考题

（1）分析本实验中半桥接法、全桥接法与 1/4 桥接法的异同。

（2）中性层实测应变不为零，可能的原因有哪些？

（3）所加载的载荷恒定时，应变沿梁高度按什么规律分布？其理论依据是什么？试用理论公式进行分析。

（4）在同一测点，应变随载荷的增加按什么规律变化？其理论依据是什么？试用理论公式进行分析。

# 实验六  弯扭组合时的主应力主方向测定

## 一、实验目的

（1）用电阻应变花测定平面应力状态下一点的主应力大小和方向。
（2）了解平面应力状态下的应变分析理论在实验中的应用。
（3）进一步熟悉使用电阻应变仪测量桥路和静态多点应变的方法。

## 二、实验设备和仪器

（1）弯扭组合实验装置。
（2）静态电阻应变仪。

## 三、实验原理

### 1. 实验装置和布片图

图 2-16 所示为弯扭组合装置简图，图 2-17 所示为弯扭组合装置的布片图。在长度为 $L$ 的薄壁圆筒横截面的上、下两测量点 $A$、$B$ 上分别贴有两枚 $45°$ 应变花，扭臂长度为 $a$，在砝码的重力作用下，薄壁圆筒产生弯扭组合变形。

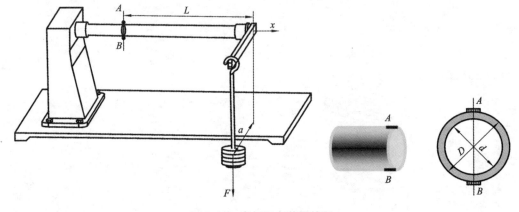

图 2-16  弯扭组合装置简图

### 2. 弯扭组合装置受力分析

本实验中，弯扭组合装置受力和受力分析如图 2-18 和图 2-19 所示。

### 3. 测点 $A$、$B$ 的应力分析

测点 $A$、$B$ 分别为薄壁圆筒的上、下应变花的粘贴点，其应力状态如图 2-20 所示。

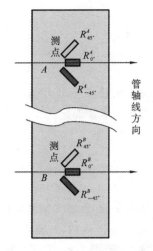 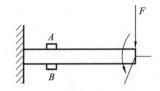

图 2-17  弯扭组合装置布片图　　　图 2-18  弯扭组合装置受力简图

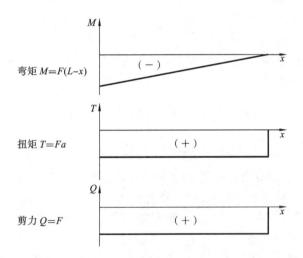

图 2-19  弯扭组合装置受力分析

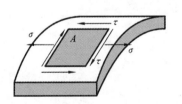 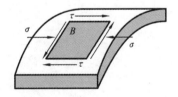

图 2-20  $A$、$B$ 点的应力状态

对于 $A$、$B$ 两点,扭矩引起的切应力大小为

$$\tau = \frac{T_a}{W_T} = \frac{Fa}{W_T}$$

弯矩引起的正应力大小为

$$\sigma = \frac{M}{W_z} = \frac{FL}{W_z}$$

由上述可知,$A$、$B$ 点的主应力理论大小为

$$\sigma_1(\sigma_3) = \sigma/2 \pm \sqrt{(\sigma/2)^2 + \tau^2}$$

$A$、$B$ 点的理论主方向为

$$\alpha = \arctan(-2\tau/\sigma)/2$$

### 4. 平面应力状态的应变分析原理

根据平面应力状态的应变分析原理,用三个不同方向的线应变即可确定一点的主应力大小和方向。平面应力状态的主应力-主应变关系由广义胡克定律确定:

$$\sigma_1 = \frac{E}{1-\mu^2}(\varepsilon_1 + \mu\varepsilon_3) \tag{1}$$

$$\sigma_3 = \frac{E}{1-\mu^2}(\varepsilon_3 + \mu\varepsilon_1) \tag{2}$$

但在平面应力状态的一般情况下,主应变的方向是未知的,所以无法直接用应变片测量主应变。根据平面应力状态的应变分析原理,在 $x$-$y$ 直角平面坐标内,一点与 $x$ 轴成 $\alpha$ 角($\alpha$ 以逆时针方向为正)的方向线应变 $\varepsilon_a$ 与该点沿 $x$、$y$ 方向的线应变 $\varepsilon_x$、$\varepsilon_y$ 和 $x$-$y$ 平面的切应变 $\gamma_{xy}$ 之间有下列关系:

$$\varepsilon_a = \frac{\varepsilon_x + \varepsilon_y}{2} + \frac{\varepsilon_x - \varepsilon_y}{2}\cos2\alpha - \frac{1}{2}\gamma_{xy}\sin2\alpha \tag{3}$$

$\varepsilon_a$ 随 $\alpha$ 角的变化而改变,在两个相互垂直的主方向上,$\varepsilon_a$ 到达极值,即为主应变。由式(3)可得两主应变的大小和方向为

$$\varepsilon_{1,3} = \frac{\varepsilon_x + \varepsilon_y}{2} \pm \frac{1}{2}\sqrt{(\varepsilon_x - \varepsilon_y)^2 + \gamma_{xy}^2} \tag{4}$$

$$\tan2\alpha_0 = -\frac{\gamma_{xy}}{\varepsilon_x - \varepsilon_y} \tag{5}$$

由于切应变 $\gamma_{xy}$ 无法用应变片测得,因此可以任意选择三个 $\alpha$ 角,测量三个方向的线应变,将它们分别代入式(3),可得三个独立方程,分别求解 $\varepsilon_x$、$\varepsilon_y$ 和 $\gamma_{xy}$,然后再把 $\varepsilon_x$、$\varepsilon_y$ 和 $\gamma_{xy}$ 代入式(4)和式(5),即可求得主应变 $\varepsilon_1$、$\varepsilon_3$ 的大小和方向,最后由式(1)、式(2)求得主应力的大小,主应力的方向和主应变方向一致。

将本实验中测得的 45°应变花的数据代入以上公式,可得出式(6)至式(8)所示的简易公式。

主方向:

$$\tan2\alpha_0 = \frac{\varepsilon_{45°} - \varepsilon_{-45°}}{2\varepsilon_{0°} - \varepsilon_{-45°} - \varepsilon_{45°}} \tag{6}$$

主应变:

$$\varepsilon_{1,3} = \frac{\varepsilon_{45°} + \varepsilon_{-45°}}{2} \pm \frac{\sqrt{2}}{2}\sqrt{(\varepsilon_{-45°} - \varepsilon_{0°})^2 + (\varepsilon_{45°} - \varepsilon_{0°})^2} \tag{7}$$

主应力:

$$\sigma_1 = \frac{E}{1-\mu^2}(\varepsilon_1 + \mu\varepsilon_3), \quad \sigma_3 = \frac{E}{1-\mu^2}(\varepsilon_3 + \mu\varepsilon_1) \tag{8}$$

## 四、实验步骤

应变仪桥路仍选 1/4 桥接线方法,砝码手动加载,每个砝码的重力为 9.8 N,每组配有 5 个等量砝码。

## 五、实验数据记录与计算

薄壁圆筒的基本尺寸和应变花参数记录于表 2-7,薄壁圆筒测点 $A$、$B$ 的应变数据记录于表 2-8。

表 2-7　薄壁圆筒测点 $A$、$B$ 的基本尺寸和应变花参数

| 薄壁圆筒外径 $D/\text{mm}$ | 薄壁圆筒内径 $d/\text{mm}$ | $L/\text{mm}$ | $a/\text{mm}$ | 泊松比 $\mu$ | 弹性模量 $E/\text{GPa}$ | 灵敏系数 $K$ | 电阻值 $R_0$ |
|---|---|---|---|---|---|---|---|
|  |  |  |  |  |  |  |  |

表 2-8　薄壁圆筒测点 $A$、$B$ 的应变数据

| 载荷/N | 测点应变($\mu\varepsilon$) | | | | | |
|---|---|---|---|---|---|---|
|  | 上点 $A$ | | | 下点 $B$ | | |
|  | $\varepsilon_{-45°}$ | $\varepsilon_{0°}$ | $\varepsilon_{45°}$ | $\varepsilon_{-45°}$ | $\varepsilon_{0°}$ | $\varepsilon_{45°}$ |
| 9.8 |  |  |  |  |  |  |
| $2\times9.8$ |  |  |  |  |  |  |
| $3\times9.8$ |  |  |  |  |  |  |
| $4\times9.8$ |  |  |  |  |  |  |
| $5\times9.8$ |  |  |  |  |  |  |
| $\overline{\varepsilon_i}=$ | $\overline{\varepsilon_{-45°}}=$ | $\overline{\varepsilon_{0°}}=$ | $\overline{\varepsilon_{45°}}=$ | $\overline{\varepsilon_{-45°}}=$ | $\overline{\varepsilon_{0°}}=$ | $\overline{\varepsilon_{45°}}=$ |

## 六、思考题

(1) 如何确定薄壁圆筒弯扭组合变形时横截面上危险点的位置及该点的应力状态?

(2) 平面应力状态下,如何通过测量某一点的三个不同方向的线应变来确定该点的主应力大小和方向?

# 实验七    应变片粘贴实验

## 一、实验目的

(1) 初步掌握应变片的粘贴技术。
(2) 学习贴片质量检查的一般方法。

## 二、实验设备和仪器

(1) 电阻应变仪。
(2) 等强度梁试样。
(3) 数字万用表。
(4) 应变片、502 快干胶、连接导线及接线端子片。
(5) 其他:砂纸、丙酮或无水酒精、药棉等清洗材料,电烙铁、镊子、直尺等工具。

## 三、应变片粘贴工艺

电测应力分析中,构件表面的应变通过黏结层传递给应变片。测量数据的可靠性很大程度上取决于应变片的粘贴质量,这就要求黏结层薄而均匀,无气泡,充分固化,既不产生蠕滑又不脱胶。应变片的粘贴全由手工操作,要保证位置准确、质量优良,必须反复实践,积累经验。应变片的粘贴工艺包括下列几个过程。

### 1. 应变片的筛选

应变片的丝栅或箔栅要排列整齐,无弯折,无锈蚀斑痕,基底不能有破损。对于经筛选后的同一批应变片,要用数字万用表逐片测量其电阻值,其电阻值相差不应超过 $0.5\ \Omega$。

### 2. 试样表面处理

为使应变片粘贴牢固,试样上粘贴应变片的部位应刮去油漆,打磨锈斑,除去油污。表面粗糙度达到 $Ra\ 20\sim25$。表面为贴片而进行处理的面积应大于应变片基底面积的三倍。若表面过于光滑,则应用细砂纸打出与应变片轴线成 $45°$ 的交叉纹路。打磨平整后,用划针沿贴片方位划出标志线。

贴片前,用药棉或纱布蘸丙酮或无水酒精清洗试样的打磨部位,直至药棉上看不见污渍为止。待丙酮或无水酒精挥发,表面干燥后,方可进行贴片。

### 3. 应变片粘贴

常用的常温应变片的黏结剂有 502(或 501)快干胶、环氧树脂胶、酚醛树脂胶等。在寒冷或潮湿的环境下,贴片前,最好用电吹风的热风将试样贴片部位加热至30~40 ℃。贴片时,在粘贴表面先涂一薄层黏结剂。用手指捏住(或镊子钳住)应变片的引出线,在基底上也涂上黏结剂,即刻放置于试样上,且使应变片基准线对准刻于试样上的标志线。盖上聚氯乙烯透明薄膜(或玻璃纸),用拇指沿应变片轴线朝一个方向滚压,手感由轻到重,挤出气泡和多余的胶水,保证黏结层尽可能薄而均匀,且避免应变片滑动或转动。按压半分钟左右,使应变片粘牢。经过适宜的干燥时间后,轻轻揭去聚氯乙烯薄膜(或玻璃纸),观察粘贴情况。如敏感栅部位有气泡,应将应变片铲除,重新清理、贴片。若敏感栅部位粘牢,只是基底边缘翘起,则主要在这些局部补充粘贴即可。

应变片粘贴完成要待黏结剂完全固化后才可使用。不同种类的黏结剂固化要求各异。502胶可自然固化。黏结剂固化前,用镊子把应变片引出线拉起,使它不与试样接触。

### 4. 导线的连接和固定

连接应变片和应变仪的导线一般可用聚氯乙烯双芯多股铜导线或丝包漆包线。导线与应变片引出线的连接最好用接线端子片作为过渡。接线端子片用502胶固定在试样上,导线头和接线端子片的铜箔都要预先挂锡,然后将应变片引出线和导线焊接在端子片上。也可把应变片引出线直接缠绕在导线上,然后上锡焊接,并在焊锡头与试样之间用涤纶绝缘胶带隔开。不论用何种方法连接都不能出现"虚焊"现象。最后,将压线片(不锈钢箔)用点焊机焊在试样上,以固定导线。也可用胶布代替压线片将导线固定在试样上。

### 5. 应变片粘贴工艺的质量检查

贴片质量的好坏是电测成败的关键,良好的贴片质量需要熟练的粘贴技术,还需要外观质量和内在质量的保证。

(1)外观质量。粘贴于构件上的应变片,胶层应薄而均匀,透过敏感栅,黏结剂应具有透明感。黏结剂太少,粘贴时滚压不当,敏感栅部位将形成气泡,胶层不均匀,黏结剂太多会造成应变片局部隆起,应变片发生折皱等都是不允许的,应将应变片铲除重贴。应变片引出线不能粘于构件上。

(2)内在质量。应变片粘贴完成后,用数字万用表测量其电阻值。贴片前后应变片的电阻应无较大变化。如有较大变化,说明粘贴时应变片有折皱,最好重贴。黏结剂固化后,用低压兆欧表测量引出线与构件间的绝缘电阻。对于用于短期测量的应变片,绝缘电阻要求为50~100 MΩ;对于用于长期测量、高湿度环境或水下环境的应变片,绝缘电阻要求在500 MΩ以上。绝缘电阻的高低是应变片粘贴质量的重要指标,绝缘电阻偏低,应变片的零漂、蠕变、滞后都较严重,将引起较大的测量误差。黏结剂未充分固化也会引起绝缘电阻偏低,可用电吹风加热以加速固化。

导线焊接后,应再一次测量应变片电阻值和绝缘电阻。由于导线有电阻,测出的电阻值略有增加是正常的。但如读数漂移不定,一般是焊接不良所致,应重新焊接。导线连接后的绝缘电阻如低于导线连接以前的值,一般是接线端子片基底被烧穿引起的,应更换接线端子片。

(3)质量的综合评定。应变片粘贴工艺质量最终应由实测时应变片的表现来评定。应变仪是高灵敏度的仪器,应变片接入应变仪后,那些通过外观检查、万用表测定都难以发现的隐患皆将暴露无遗。诸如,由于电阻值变化太大电桥无法平衡;由虚焊或绝缘电阻过低产生的漂移;由于气泡等因素,当以橡皮、软物件轻压应变片敏感栅时,应变指示变化较大等。这些缺陷都应在正式测量之前,采取措施消除。

### 6. 应变片的防潮保护

粘贴好的应变片,如长期暴露于空气中,会因受潮而使黏结牢度降低,绝缘电阻减小,严重的会造成应变片剥离脱落,因此应敷设防潮保护层。

防潮保护层涂敷之前,可把涂敷部位加热至 40～50 ℃,以保证黏结良好。保护层厚 1～2 mm,周边超出应变片 10～20 mm,最好将焊锡头及接线端子片等都埋入防潮保护层中。

## 四、实验步骤

(1)每组应变片 4 枚,等强度梁试样 2 根。

(2)试样清理。

(3)贴片。检查应变片后,按照贴片的工艺要求,沿等强度梁的纵向和横向各贴一枚应变片,具体如图 2-21 所示。

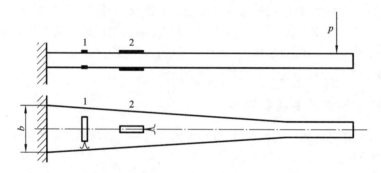

**图 2-21    等强度梁应变片的粘贴示意图**

(4)焊接。焊接完毕后,检查焊接质量,用万用表测量 $R_0$ 值,并记录。

## 五、注意事项

(1) 502 快干胶黏结力很强,且有强烈的刺激异味,应避免过量吸收,如皮肤或衣物被粘住,应以丙酮或无水酒精清洗,不要用力拉扯。

(2)应变片引出线与敏感栅的焊点很脆弱,不要拉脱。

## 六、思考题

(1)检查贴片质量时,是否可省略外观质量检查和内在质量检查这两个程序?

(2)粘贴的用于测量的纵向和横向应变片没有附着在同一横截面上,并且远离固定端。这对测量结果有影响吗?如果将其放置在尽可能靠近固定端的位置,结果是否准确?

# 实验八 等强度梁的测试实验

## 一、实验目的

(1) 熟练使用静态电阻应变仪,进一步训练电阻应变测量技术中的组桥技巧。
(2) 测定等强度梁上的应变,验证等强度梁的应力分布。
(3) 测算等强度梁所用材料的弹性模量 $E$ 及泊松比 $\mu$。
(4) 用电测法测量待测重物的重力 $P$。

## 二、实验设备和仪器

(1) 等强度悬臂梁实验装置。
(2) 静态电阻应变仪。
(3) 直尺及游标卡尺。

## 三、实验内容与要求

(1) 根据实验目的,思考并拟订实验方案。
(2) 表格设计。测量梁的几何尺寸($L$、$a$、$b_0$、$b_1$ 和 $h$),设计应变片粘贴方案、接桥方式、加载、应变仪读数等,填入相应表格中。
(3) 测算悬臂梁所用材料的弹性模量 $E$ 和泊松比 $\mu$,并进行材料对比。
(4) 测量待测重物的重力 $P$。

## 四、实验步骤

(1) 每组应变片 4 枚,等强度梁试样 2 根。
(2) 设计贴片方案,画出贴片图,绘制数据表格。
(3) 试样清理后,按照贴片的工艺要求进行贴片和焊接。
(4) 组桥接入应变仪。加载,记录数据。

## 五、等强度梁装置

等强度梁装置如图 2-22 和图 2-23 所示。

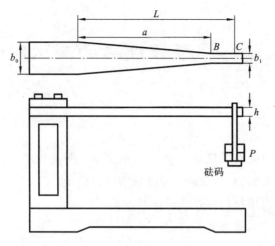

图 2-22　等强度梁实验装置

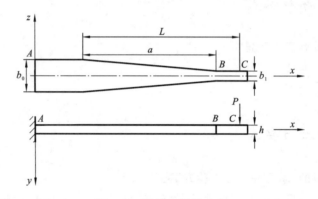

图 2-23　等强度梁的尺寸

## 六、思考题

(1) 什么是等强度梁?

(2) 等强度梁表面的应力如何计算?

(3) 等强度梁的弹性模量和泊松比如何计算?

(4) 如采用 1/4 桥、半桥或全桥测量,应如何布置电阻应变片,如何组桥?

(5) 为测定 $\mu$ 而粘贴的纵、横两枚应变片未贴在同一横截面位置上,又离固定端较远,这对测量结果有无影响? 如使其尽量靠近固定端,结果会更准确吗?

# 实验九  梁的冲击载荷系数测定

在工程实践中,除了静载荷外,经常会遇到动载荷问题。在动载荷作用下,构件各点的应力、应变与受静载荷作用时有很大不同。按照加载速度的不同,动载荷也有不同的形式。在极短的时间内以很大的速度作用在构件上的载荷,称为冲击载荷,它是一种常见的动载荷形式。冲击载荷作用在构件上时产生的应力称为冲击应力。因此对于承受冲击力的构件(如锻造、冲孔、凿岩等)来说,冲击应力是设计中应考虑的主要问题。

## 一、实验目的

(1) 运用实验的方法测定冲击应力及动载荷系数。
(2) 了解动应力的测量原理、方法及仪器。

## 二、实验设备和仪器

(1) NI 信号采集系统(NIcDAQ-9178、NI9237 及 NI9949)。
(2) 等强度梁实验装置。
(3) 游标卡尺及卷尺。

## 三、实验原理

本实验采用的变截面等强度梁如图 2-22 所示,其在砝码端受到砝码在高度 $H$ 处自由落下的冲击作用。由理论可知该梁的动载荷系数为

$$K_{d}=1+\sqrt{1+\frac{2H}{\delta_{j}}}$$

式中:$H$ 为砝码自由落下的高度;$\delta_{j}$ 为梁的静挠度。梁的尺寸为:$L=140$ mm,$b_{0}=30$ mm,$h=4$ mm。等强度梁简化模型如图 2-24 所示。

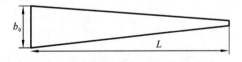

**图 2-24  等强度梁简化模型**

砝码从高度 $H$ 处落下冲击等强度梁时,测点的动应变 $\varepsilon_{d}$ 将通过 NI 采集卡记录下来。再将砝码静止放在梁上可测得同一点的静应变 $\varepsilon_{j}$。动载荷系数为

$$K_{d}=\varepsilon_{d}/\varepsilon_{j}$$

冲击应力为

$$\sigma_{d}=E\varepsilon_{d}\quad 或\quad \sigma_{d}=EK_{d}\varepsilon_{j}$$

式中:$E$ 为梁的弹性模量(该等强度梁的材质为铝合金,约 68 GPa)。

## 四、实验步骤

(1) 测量并记录简支梁的几何尺寸、重物高度和重力及材料的弹性模量。

(2) 连接导线:按 1/4 桥方法接入 NI 信号采集系统。

(3) 打开自编的专用软件 SIGNAL EXPRESS。

(4) 设置参数,进行应变标定。记录砝码高度,释放砝码,采样,记录最大动应变,波形示意图如图 2-25 所示。

(5) 重复第(4)步,平稳放置砝码,采集静应变。

(6) 计算最大冲击应力及实测动载荷系数,考察其与高度的关系。

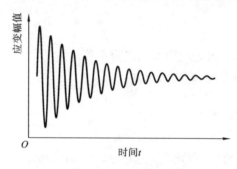

**图 2-25　波形示意图**

## 五、预习要求

(1) 复习冲击动载荷系数的概念及计算方法。

(2) 了解动应变测量方法及动应变标定方法。

## 六、实验报告要求

实验报告应包括:实验名称、实验目的、实验装置草图、仪器名称和规格、原始数据及实验结果等。实验结果应包括:数据记录、计算结果及曲线绘制等。

# 材料力学实验报告

## 实验一　金属材料的拉伸实验

一、实验目的

二、实验设备和仪器

三、实验原理

## 四、实验步骤

## 五、实验数据记录与计算

计算精确度：

(1) 强度性能指标(下屈服强度 $\sigma_{eL}$ 和抗拉强度 $\sigma_m$)的计算精度要求为 0.5 MPa,即舍去所有小于 0.25 MPa 的数值,大于或等于 0.25 MPa 而小于 0.75 MPa 的数值化为 0.5 MPa,大于或等于 0.75 MPa 的数值则进为 1 MPa。

(2) 塑性性能指标(断后伸长率 $A$ 和断面收缩率 $Z$)的计算精度要求为 0.5%,即舍去所有小于0.25%的数值,大于或等于 0.25% 而小于 0.75% 的数值化为 0.5%,大于或等于 0.75%的数值则进为 1%。

表 1    低碳钢拉伸的实验数据

| 实验数据 | | | 第一次测量 | 第二次测量 | 平均值 |
|---|---|---|---|---|---|
| 拉伸前 | 截面一直径 | $d_{01}/mm$ | | | |
| | 截面二直径 | $d_{02}/mm$ | | | |
| | 截面三直径 | $d_{03}/mm$ | | | |
| | 原始标距 | $L_0/mm$ | | | |
| 拉伸后 | 最小截面直径 | $d_1/mm$ | | | |
| | 断后标距 | $L_u/mm$ | | | |
| | 下屈服载荷/kN | $F_{eL}$ | | | |
| | 最大载荷/kN | $F_m$ | | | |

$\sigma_{eL} = F_{eL}/S_0 =$

$\sigma_m = F_m/S_0 =$

$A = (L_u - L_0)/L_0 \times 100\% =$

$Z = (S_0 - S_u)/S_0 \times 100\% =$

表 2    灰铸铁拉伸的实验数据

| 实验数据 | | | 第一次测量 | 第二次测量 | 平均值 |
|---|---|---|---|---|---|
| 拉伸前 | 截面一直径 | $d_{01}/mm$ | | | |
| | 截面二直径 | $d_{02}/mm$ | | | |

| 实验数据 | | | 第一次测量 | 第二次测量 | 平均值 |
|---|---|---|---|---|---|
| 拉伸前 | 截面三直径 | $d_{03}$/mm | | | |
| | 原始标距 | $L_0$/mm | | | |
| 拉伸后 | 最大载荷/kN | $F_m$ | | | |

$\sigma_m = F_m/S_0 =$

## 六、实验结果

画出低碳钢和灰铸铁的拉伸曲线简图,并标注关键点。

## 七、思考题

(1) 低碳钢和灰铸铁在常温静载拉伸时的力学性能和破坏形式有何异同?

(2) 材料和直径相同但长度不同的试样,其伸长率是否相同? 为什么?

(3) 你认为使实验结果产生误差的因素有哪些? 应如何避免或减小其影响?

(4) 测定材料的力学性能有何实用价值?

# 实验二 金属材料的扭转实验

## 一、实验目的

## 二、实验设备和仪器

## 三、实验原理

## 四、实验数据记录与计算

表 1 低碳钢扭转的数据

| 低碳钢实验数据 | | 第一次测量 | 第二次测量 | 平均值 |
|---|---|---|---|---|
| 截面一直径 | $d_{01}/\text{mm}$ | | | |
| 截面二直径 | $d_{02}/\text{mm}$ | | | |
| 截面三直径 | $d_{03}/\text{mm}$ | | | |
| 屈服扭矩/(N·m) | $T_{eL}$ | | | |
| 最大扭矩/(N·m) | $T_m$ | | | |

表 2　灰铸铁扭转的数据

| 灰铸铁实验数据 | | 第一次测量 | 第二次测量 | 平均值 |
|---|---|---|---|---|
| 截面一直径 | $d_{01}/mm$ | | | |
| 截面二直径 | $d_{02}/mm$ | | | |
| 截面三直径 | $d_{03}/mm$ | | | |
| 最大扭矩/(N·m) | $T_m$ | | | |

低碳钢：

$\tau_{eL} = 0.75 T_{eL}/W =$

$\tau_m = 0.75 T_m/W =$

灰铸铁：

$\tau_m = T_m/W =$

# 五、思考题

（1）画出低碳钢和灰铸铁的扭转曲线图，并在图上标出关键点。

（2）画出低碳钢和灰铸铁的断口示意图。

（3）比较低碳钢与灰铸铁试样的扭转破坏断口，并分析它们的破坏原因。

（4）分析比较低碳钢和灰铸铁在拉伸和扭转时的变形情况和破坏特点，归纳这两种材料的力学性能。

# 实验三　电测法基础及应变仪使用

## 一、实验原理

(1) 什么是电测法？电测法有什么优缺点？

(2) 什么是电阻应变效应？

(3) 什么是应变电桥？简述应变电桥的原理。

(4) 应变片在电桥中的接法有几种？请画出相应的桥路原理图。

(5) 什么是温度效应？排除温度效应的措施是什么？

## 二、实验步骤

应变片接线实验：以等截面梁的弯曲正应力实验装置为例，进行半桥单臂（1/4 桥）接法、半桥双臂接法和全桥接法练习。请画出应变片的桥路接线图，并写出应变公式。

（1）半桥单臂接法。

（2）半桥双臂接法。

（3）全桥接法。

# 实验四　矩形截面梁弯曲正应力测定实验(1/4桥)

## 一、实验目的

## 二、实验设备和仪器

## 三、实验原理

## 四、实验步骤

## 五、实验数据记录与计算

表 1　矩形截面梁的参数

| $a$/mm | $b$/mm | $c$/mm | $L$/mm | $h$/mm | $E$/GPa | $K$ | $R_0$ |
|--------|--------|--------|--------|--------|---------|-----|-------|
|        |        |        |        |        |         |     |       |

表 2　矩形截面梁纯弯曲实验数据

| 载荷/kN | 测点应变($\mu\varepsilon$) | | | | | | | |
|---|---|---|---|---|---|---|---|---|
| | $y_1 = 0$ | $y_2 = 0$ | $y_3 = h/4$ | $y_4 = h/4$ | $y_5 = -h/4$ | $y_6 = -h/4$ | $y_7 = h/2$ | $y_8 = -h/2$ |
| 1 | | | | | | | | |
| 1.5 | | | | | | | | |
| 2 | | | | | | | | |
| 2.5 | | | | | | | | |
| 3 | | | | | | | | |
| $\overline{\varepsilon_i} =$ | $\overline{\varepsilon_1} =$ | $\overline{\varepsilon_2} =$ | $\overline{\varepsilon_3} =$ | $\overline{\varepsilon_4} =$ | $\overline{\varepsilon_5} =$ | $\overline{\varepsilon_6} =$ | $\overline{\varepsilon_7} =$ | $\overline{\varepsilon_8} =$ |
| $\sigma_{实} = E\overline{\varepsilon}$ | | | | | | | | |
| $\sigma_{理} = \dfrac{\overline{M}y}{I_z}$ | | | | | | | | |
| 误差 | | | | | | | | |

(1) $\overline{\varepsilon_1} =$

$\sigma_{实1} = E\overline{\varepsilon_1}$

$\sigma_{理1} = \dfrac{\overline{M}y}{I_z}$

相对误差(%):

(2) $\overline{\varepsilon_2} =$

$\sigma_{实2} = E\overline{\varepsilon_2}$

$\sigma_{理2} = \dfrac{\overline{M}y}{I_z}$

相对误差(%):

(3) $\overline{\varepsilon_3} =$

$\sigma_{实3} = E\overline{\varepsilon_3}$

$\sigma_{理3} = \dfrac{\overline{M}y}{I_z}$

相对误差(%):

(4) $\overline{\varepsilon_4} =$

$\sigma_{实4} = E\overline{\varepsilon_4}$

$\sigma_{理4} = \dfrac{\overline{M}y}{I_z}$

相对误差(%):

(5) $\overline{\varepsilon_5} =$

$\sigma_{实5} = E\overline{\varepsilon_5}$

$\sigma_{理5} = \dfrac{\overline{M}y}{I_z}$

相对误差(%):

(6) $\overline{\varepsilon_6} =$

$\sigma_{实6} = E\overline{\varepsilon_6}$

$$\sigma_{理6} = \frac{\overline{M}y}{I_z}$$

相对误差(%):

(7) $\overline{\varepsilon_7} =$

$$\sigma_{实7} = E\overline{\varepsilon_7}$$

$$\sigma_{理7} = \frac{\overline{M}y}{I_z}$$

相对误差(%):

(8) $\overline{\varepsilon_8} =$

$$\sigma_{实8} = E\overline{\varepsilon_8}$$

$$\sigma_{理8} = \frac{\overline{M}y}{I_z}$$

相对误差(%):

表3 矩形截面梁横力弯曲实验数据

| 载荷/kN | 测点应变($\mu\varepsilon$) | | | | | | | |
|---|---|---|---|---|---|---|---|---|
| | $y_1 = 0$ | $y_2 = 0$ | $y_3 = h/4$ | $y_4 = h/4$ | $y_5 = -h/4$ | $y_6 = -h/4$ | $y_7 = h/2$ | $y_8 = -h/2$ |
| 1 | | | | | | | | |
| 1.5 | | | | | | | | |
| 2 | | | | | | | | |
| 2.5 | | | | | | | | |
| 3 | | | | | | | | |
| $\overline{\varepsilon_i} =$ | $\overline{\varepsilon_1} =$ | $\overline{\varepsilon_2} =$ | $\overline{\varepsilon_3} =$ | $\overline{\varepsilon_4} =$ | $\overline{\varepsilon_5} =$ | $\overline{\varepsilon_6} =$ | $\overline{\varepsilon_7} =$ | $\overline{\varepsilon_8} =$ |
| $\sigma_{实} = E\overline{\varepsilon}$ | | | | | | | | |
| $\sigma_{理} = \frac{\overline{M}y}{I_z}$ | | | | | | | | |
| 误差 | | | | | | | | |

(1) $\overline{\varepsilon_1} =$

$$\sigma_{实1} = E\overline{\varepsilon_1}$$

$$\sigma_{理1} = \frac{\overline{M}y}{I_z}$$

相对误差(%):

(2) $\overline{\varepsilon_2} =$

$$\sigma_{实2} = E\overline{\varepsilon_2}$$

$$\sigma_{理2} = \frac{\overline{M}y}{I_z}$$

相对误差(%):

(3) $\overline{\varepsilon_3} =$

$$\sigma_{实3} = E\overline{\varepsilon_3}$$

$$\sigma_{理3}=\frac{\overline{M}y}{I_z}$$

相对误差(%)：

(4) $\overline{\varepsilon_4}=$

$$\sigma_{实4}=E\overline{\varepsilon_4}$$

$$\sigma_{理4}=\frac{\overline{M}y}{I_z}$$

相对误差(%)：

(5) $\overline{\varepsilon_5}=$

$$\sigma_{实5}=E\overline{\varepsilon_5}$$

$$\sigma_{理5}=\frac{\overline{M}y}{I_z}$$

相对误差(%)：

(6) $\overline{\varepsilon_6}=$

$$\sigma_{实6}=E\overline{\varepsilon_6}$$

$$\sigma_{理6}=\frac{\overline{M}y}{I_z}$$

相对误差(%)：

(7) $\overline{\varepsilon_7}=$

$$\sigma_{实7}=E\overline{\varepsilon_7}$$

$$\sigma_{理7}=\frac{\overline{M}y}{I_z}$$

相对误差(%)：

(8) $\overline{\varepsilon_8}=$

$$\sigma_{实8}=E\overline{\varepsilon_8}$$

$$\sigma_{理8}=\frac{\overline{M}y}{I_z}$$

相对误差(%)：

## 六、思考题

(1) 载荷一定时,应变沿梁高度按什么规律分布？其理论依据是什么？写出公式。

(2) 在同一测点,应变随载荷增加按什么规律变化？其理论依据是什么？写出公式。

# 实验五　矩形截面梁弯曲正应力测定实验(半桥与全桥)

## 一、实验目的

## 二、实验设备和仪器

## 三、实验原理

## 四、实验步骤

## 五、实验数据记录与计算

表1　矩形截面梁参数

| $a/\mathrm{mm}$ | $b/\mathrm{mm}$ | $c/\mathrm{mm}$ | $L/\mathrm{mm}$ | $h/\mathrm{mm}$ | $E/\mathrm{GPa}$ | $K$ | $R_0$ |
|---|---|---|---|---|---|---|---|
|  |  |  |  |  |  |  |  |

表 2 矩形截面梁的半桥实验数据

| 载荷/kN | 测点应变($\mu\varepsilon$) | | | | | |
|---|---|---|---|---|---|---|
| | 纯弯曲 | | | 横力弯曲 | | |
| | $y_1 =$ | $y_2 =$ | $y_3 =$ | $y_4 =$ | $y_5 =$ | $y_6 =$ |
| 1 | | | | | | |
| 1.5 | | | | | | |
| 2 | | | | | | |
| 2.5 | | | | | | |
| 3 | | | | | | |
| $\overline{\varepsilon_i} =$ | $\overline{\varepsilon_1} =$ | $\overline{\varepsilon_2} =$ | $\overline{\varepsilon_3} =$ | $\overline{\varepsilon_4} =$ | $\overline{\varepsilon_5} =$ | $\overline{\varepsilon_6} =$ |
| $\sigma_{实} = E\overline{\varepsilon}/2$ | | | | | | |
| $\sigma_{理} = \dfrac{\overline{M}y}{I_z}$ | | | | | | |
| 误差 | | | | | | |

纯弯曲：

(1) $\overline{\varepsilon_1} =$

$\sigma_{实1} = E\overline{\varepsilon_1}/2$

$\sigma_{理1} = \dfrac{\overline{M}y}{I_z}$

相对误差(%)：

(2) $\overline{\varepsilon_2} =$

$\sigma_{实2} = E\overline{\varepsilon_2}/2$

$\sigma_{理2} = \dfrac{\overline{M}y}{I_z}$

相对误差(%)：

(3) $\overline{\varepsilon_3} =$

$\sigma_{实3} = E\overline{\varepsilon_3}/2$

$\sigma_{理3} = \dfrac{\overline{M}y}{I_z}$

相对误差(%)：

横力弯曲：

(4) $\overline{\varepsilon_4} =$

$\sigma_{实4} = E\overline{\varepsilon_4}/2$

$\sigma_{理4} = \dfrac{\overline{M}y}{I_z}$

相对误差(%)：

(5) $\overline{\varepsilon_5}=$

$\sigma_{\text{实}5}=E\overline{\varepsilon_5}/2$

$\sigma_{\text{理}5}=\dfrac{\overline{M}y}{I_z}$

相对误差(%):

(6) $\overline{\varepsilon_6}=$

$\sigma_{\text{实}6}=E\overline{\varepsilon_6}/2$

$\sigma_{\text{理}6}=\dfrac{\overline{M}y}{I_z}$

相对误差(%):

表 3　矩形截面梁的全桥实验数据

| 载荷/kN | 测点应变($\mu\varepsilon$) | |
|---|---|---|
| | 纯弯曲 | 横力弯曲 |
| | $y_1=$ | $y_2=$ |
| 1 | | |
| 1.5 | | |
| 2 | | |
| 2.5 | | |
| 3 | | |
| $\overline{\varepsilon_i}=$ | $\overline{\varepsilon_1}=$ | $\overline{\varepsilon_2}=$ |
| $\sigma_{\text{实}}=E\overline{\varepsilon}/4$ | | |
| $\sigma_{\text{理}}=\dfrac{\overline{M}y}{I_z}$ | | |
| 误差(%) | | |

纯弯曲:

(1) $\overline{\varepsilon_1}=$

$\sigma_{\text{实}1}=E\overline{\varepsilon_1}/4$

$\sigma_{\text{理}1}=\dfrac{\overline{M}y}{I_z}$

相对误差(%):

横力弯曲:

(2) $\overline{\varepsilon_2}=$

$\sigma_{\text{实}2}=E\overline{\varepsilon_2}/4$

$\sigma_{\text{理}2}=\dfrac{\overline{M}y}{I_z}$

相对误差(%):

## 六、思考题

（1）中性层实测应变不为零的原因可能是什么？试用相邻测量点的应变值进行分析。

（2）分析本实验中半桥、全桥接法与 1/4 桥接法的异同。

（3）影响实验结果的主要因素是什么？

（4）弯曲正应力的大小是否会受材料弹性模量的影响？为什么？

## 实验六　弯扭组合时的主应力主方向测定

### 一、实验目的

### 二、实验设备和仪器

### 三、实验原理

(1) 画出实验装置简图及受力分析图。

(2) 简述平面应力状态的应变分析原理。

### 四、实验步骤

## 五、实验数据记录与计算

表1　薄壁圆筒的基本尺寸和应变花参数

| 薄壁圆筒外径 $D/\text{mm}$ | 薄壁圆筒内径 $d/\text{mm}$ | $L/\text{mm}$ | $a/\text{mm}$ | 泊松比 $\mu$ | 弹性模量 $E/\text{GPa}$ | 灵敏系数 $K$ | 电阻值 $R_0$ |
|---|---|---|---|---|---|---|---|
| | | | | | | | |

注意：薄壁圆筒加工后实际尺寸有所不同，实际数据见每组仪器的标注。

表2　薄壁圆筒测点 $A$、$B$ 的应变数据

| 载荷/N | 测点应变($\mu\varepsilon$) | | | | | |
|---|---|---|---|---|---|---|
| | 上点 $A$ | | | 下点 $B$ | | |
| | $\varepsilon_{-45°}$ | $\varepsilon_{0°}$ | $\varepsilon_{45°}$ | $\varepsilon_{-45°}$ | $\varepsilon_{0°}$ | $\varepsilon_{45°}$ |
| 9.8 | | | | | | |
| $2\times9.8$ | | | | | | |
| $3\times9.8$ | | | | | | |
| $4\times9.8$ | | | | | | |
| $5\times9.8$ | | | | | | |
| $\overline{\varepsilon_i}=$ | $\overline{\varepsilon_{-45°}}=$ | $\overline{\varepsilon_{0°}}=$ | $\overline{\varepsilon_{45°}}=$ | $\overline{\varepsilon_{-45°}}=$ | $\overline{\varepsilon_{0°}}=$ | $\overline{\varepsilon_{45°}}=$ |

(1) 上点 $A$：

$\overline{\varepsilon_{-45°}}=$

$\overline{\varepsilon_{0°}}=$

$\overline{\varepsilon_{45°}}=$

主方向：$\tan2\alpha_0=\dfrac{\varepsilon_{45°}-\varepsilon_{-45°}}{2\varepsilon_{0°}-\varepsilon_{-45°}-\varepsilon_{45°}}=$

$\alpha_0=$

主应变：$\varepsilon_{1,3}=\dfrac{\varepsilon_{45°}+\varepsilon_{-45°}}{2}\pm\dfrac{\sqrt{2}}{2}\sqrt{(\varepsilon_{-45°}-\varepsilon_{0°})^2+(\varepsilon_{45°}-\varepsilon_{0°})^2}=$

$\varepsilon_1=$

$\varepsilon_3=$

主应力：$\sigma_1=\dfrac{E}{1-\mu^2}(\varepsilon_1+\mu\varepsilon_3)=$

$\sigma_3=\dfrac{E}{1-\mu^2}(\varepsilon_3+\mu\varepsilon_1)=$

(2) 下点 $B$：

$\overline{\varepsilon_{-45°}}=$

$\overline{\varepsilon_{0°}}=$

$\overline{\varepsilon_{45°}}=$

主方向:$\tan 2\alpha_0 = \dfrac{\varepsilon_{45°} - \varepsilon_{-45°}}{2\varepsilon_{0°} - \varepsilon_{-45°} - \varepsilon_{45°}} =$

$\alpha_0 =$

主应变:$\varepsilon_{1,3} = \dfrac{\varepsilon_{45°} + \varepsilon_{-45°}}{2} \pm \dfrac{\sqrt{2}}{2}\sqrt{(\varepsilon_{-45°} - \varepsilon_{0°})^2 + (\varepsilon_{45°} - \varepsilon_{0°})^2} =$

$\varepsilon_1 =$

$\varepsilon_3 =$

主应力:$\sigma_1 = \dfrac{E}{1 - \mu^2}(\varepsilon_1 + \mu\varepsilon_3) =$

$\sigma_3 = \dfrac{E}{1 - \mu^2}(\varepsilon_3 + \mu\varepsilon_1) =$

## 六、思考题

(1) 上点 $A$ 和下点 $B$ 的主应力和主方向是否相同?为什么?请画图说明。

(2) 请计算上点 $A$ 的理论主应力和主方向,求实验主应力主方向与理论主应力主方向的误差,并分析引起误差的因素有哪些。

# 实验七　应变片粘贴实验

## 一、实验目的

## 二、实验设备和仪器

## 三、实验原理

## 四、实验步骤

(1) 画出实验装置图及贴片图。

(2) 测量应变片粘贴前后的电阻值。

(3) 如何保证应变片粘贴的质量？

# 实验八　等强度梁的测试实验

## 一、实验目的

## 二、实验设备和仪器

## 三、实验方案

(1) 写出等强度梁的 $E$ 与 $\mu$ 的计算公式。

(2) 设计应变桥路。

(3) 画出贴片图。

## 四、实验数据记录与计算

（1）画出等强度梁的外形尺寸图。

（2）设计表格并记录实验数据。

（3）计算等强度梁的 $E$ 与 $\mu$，并查阅资料，判断等强度梁的材料。

（4）用电测法测量待测重物的重力。

# 实验九　梁的冲击载荷系数测定

## 一、实验目的

## 二、实验设备和仪器

## 三、实验原理

## 四、实验装置草图

## 五、实验步骤

## 六、实验结果记录

## 七、实验分析

# Introduction

## I. Main Content of This Course

Laboratory Experiments of Mechanics of Materials is a key part for the study of Mechanics of Materials. The experiments can not only help the students to consolidate and to deepen the theoretical knowledge of the mechanic courses, to study the ways to use the experiment instruments, to grasp the mechanical properties of materials and the measuring method of the strain and stress of the components, but also help the students to cultivate the ability to practice, to analyze and to handle the experiment data, and help to build a serious and careful attitude.

The experiments can be divided into the following three types.

### 1. To Measure the Mechanical Properties of Materials

The mechanical properties of materials are the indexes shown in deformation, strength, etc. , when the material is suffering a force load, for example, elastic limit, yield strength, elastic modulus, fatigue strength, impact toughness etc. These strength indexes are the basis to calculate the strength, the stiffness and the stability of the components. By performing the tensile and torsion experiments, the students can strength the knowledge related with mechanics of materials and master the basic measuring method for characteristic values of the materials. The measurement of the mechanical properties of materials should be performed according to the national standards or industrial standards.

### 2. To Validate the Basic Theories

It's a frequently used method to model the real problem to an idealized model, and to deduce generalized equations according to the scientific assumptions. For example, the bending theory for the rods is based on the plane assumption. But whether the assumptions and the simplifications are correct or not, and whether the theories can be used in real applications or not, they should be validated by experiments. The normal stress experiment of a beam in bending is of this type. New-established theories and equations are needed to be validated by experiments, which are necessary tools to validate, to adjust and to develop the theories. Also, the students can consolidate and deepen the theoretical knowledge learned in classes by performing experiments.

### 3. To Measure the Stress and the Deformation of a Component

The shape, loads, surroundings of many real components in real application are usually very complicated, e. g. the under pan of a vehicle, a dam, a hydraulic press, a structure of an airplane. The theoretical calculation of the stresses and the deformations of these components is difficult. Though the finite element method and the computer is widely used, which are very powerful tools for computing, the expected results cannot be obtained for some cases. The experimental stress analysis is a method to solve the problem of the stress analysis by experiments and there are many methods of this type. In this course, we focus on the training of the electrical measurement method to measure the strain and stress. By performing the electrical measurement experiment, the students can have a preliminary grasp of the use of electrical resistance strain indicator and understand the method of experiment measurement, improving the ability to solve real problems.

### II. Rules and Requirements of This Course

All of the students are expected to follow the requirements in order to improve the effect of the experiments and to make sure the class be performed smoothly.

(1) Preview the content carefully and finish the experiment reports according to the requirements. The objectives, principles should be understood, the instruments and the methods used in the experiments should be clear.

(2) The Laboratory Rules should be strictly obeyed. Be quiet and don't touch or use the machines or instruments that are not related with the current experiment.

(3) Be serious and careful about the experiment. Each member of the group should be given a clear assignment, with close and mutual coordination. The measured data should be practical and realistic.

(4) Take good care of all the machines and instruments. All the machines and instruments are of high complexity and precision. The students are expected to read the instructions carefully so as to make sure the normal use and precisions. The experiment should be carried out in strict accordance with the rules. Safety first, and report to the teachers in case of any faults.

(5) Experiment records should be checked after the experiments. One can leave the laboratory only after the teachers have reviewed the experiment records.

### III. Instructions for the Experiment Reports

The experiment reports are the summaries of the experiment files, attention should be paid to the followings:

(1) Mind the units of the parameters during the measurement. Besides, the precision of the instruments should be noted. The minimum value usually indicates the precision of the instruments. The measured data may be not the same after multi-metering, and the average

value should be treated as the final measured value.

(2) The conclusions in the reports should be the analysis, summaries of the main phenomena seen in the experiments and the results.

# Chapter 1

# Mechanical Property Experiment of Materials

Mechanical property experiment of materials is a widely used experiment in engineering, which provides mechanical performance parameters of materials for mechanical manufacturing, civil engineering, metallurgical engineering and other various industries, to help engineers use materials properly, and meanwhile, to ensure the safety of the machine (structure) as well as its parts (components).

## Experiment 1   Tensile Experiment of Metallic Materials

### Ⅰ. Objectives

(1) To measure the tensile strength and plasticity of the low carbon steel: lower yield strength $\sigma_{eL}$, tensile strength $\sigma_m$, percentage elongation after fracture $A$ and percentage reduction of area $Z$.

(2) To measure the tensile strength $\sigma_m$ of the gray cast iron.

(3) To compare the mechanical properties and failure modes of the low carbon steel and the gray cast iron.

### Ⅱ. Experiment Equipment and Instruments

(1) An electronic universal testing machine.

(2) A vernier caliper.

### Ⅲ. Specimen

According to the National Standard GB/T 228. 1-2010 *Metallic materials—Tensile testing—Part* 1: *Method of test at room temperature*, the cross-sections of metallic tensile experiment specimens can be divided into circular section, rectangular section, polygon section, ring section, and some other shapes for some special cases according to varieties and specifications of products, and experiment purposes. Among them, the specimens with circular cross-sections or with rectangular cross-sections are the most commonly used ones.

As shown in Figure 1-1, both specimens with circular and rectangular cross-sections are composed of three parts: parallel, transition and clamping. The length of the parallel part of the specimen is called the parallel length $L_c$, and the parallel length is the distance between the two clamps for unprocessed ones. If the relationship between the original gauge length $L_0$ and the cross-sectional area $A$ of the specimen satisfies $L_0 = k \sqrt{S_0}$, then the specimen is called the proportional specimen, otherwise, called the non-proportional one. Usually, $k$ is set to be 5.65, and the gauge length should be equal to or bigger than 15 mm. $k$ can be set to 11.3 if the area of the cross-section of the specimen is too small, or the non-proportional specimen can be used. The transition part is smoothly connected with the parallel part by an arc, so as to ensure that the fracture of the specimen is in the parallel part. The clamping part is slightly larger, and its shape and size are designed according to the specimen size, material characteristics, experiment purpose and fixture structure of the universal testing machine. Please refer to the National Standard GB/T 228.1-2010 for the technical requirements for the shape, size and processing of the specimens.

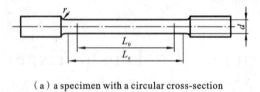

(a) a specimen with a circular cross-section

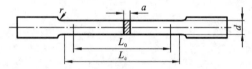

(b) a specimen with a rectangular cross-section

Figure 1-1    Specimens for the Tensile Experiment

## Ⅳ. Experiment Principles

### 1. The Tensile Experiment of the Low Carbon Steel

The low carbon steel is a typical plastic material, its tensile process can be divided into four stages: elastic stage, yield stage, strengthening stage and necking stage. The elastic modulus $E$, the lower yield load $F_{eL}$ and the maximum load $F_m$ can be measured by Figure 1-2.

Where, $L_0$ is the original gauge length of the specimen; $L_u$ is the final gauge length after fracture, $S_0$ is the original cross-sectional area of the specimen; $S_u$ is the minimum cross-sectional area of the specimen after fracture. The plastic deformation occurs at the necking position of the specimen, decreasing gradually to the two ends. Thus, the plastic elongation for the gauge length varies as the different fracture position. If the fracture position locates at the middle of the specimen, the necking stage with severe plastic deformation will be totally covered by the gauge length, and the gauge length will have a large plastic elongation; however, if the fracture position is near to either end of the specimen, only part of the neck-

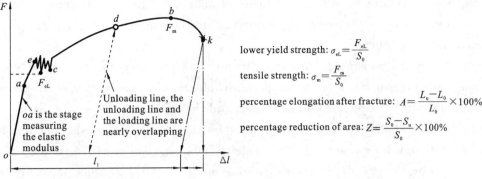

lower yield strength: $\sigma_{eL} = \dfrac{F_{eL}}{S_0}$

tensile strength: $\sigma_m = \dfrac{F_m}{S_0}$

percentage elongation after fracture: $A = \dfrac{L_u - L_0}{L_0} \times 100\%$

percentage reduction of area: $Z = \dfrac{S_0 - S_u}{S_0} \times 100\%$

Note: mind the meanings of lines $oe(oa,ae)$, $ec$, $cb$, and $bk$.

**Figure 1-2    Tensile Curve for the Low Carbon Steel**

ing stage with severe plastic deformation is covered by the gauge length in this case, the gauge length will have a smaller plastic elongation. Therefore, the fracture position has an impact on the percentage elongation measured. In order to avoid this effect, the National Standard has made the following provisions for the measurement of the final gauge length after fracture $L_u$: during the measurement, the two segments should be closely connected at the fracture point, so that the axis of the two should be in a straight line. If a gap is formed at the fracture, the gap shall be included in $L_u$. If the fracture is outside the gauge length or within the length but less than $2d$, then the test is invalid.

Determination of elastic modulus $E$: the stress is proportional to the strain within the tensile elastic range of the material,

$$\sigma_p = E \cdot \varepsilon_p \rightarrow E = \frac{\Delta F_p \cdot L_0}{S_0 \cdot \Delta L_p}.$$

## 2. The Tensile Experiment of the Gray Cast Iron

The tensile process of the gray cast iron is relatively simple, and it can be approximated that it passes through the elastic stage to fracture directly, without yield and necking stage, as shown in Figure 1-3. The percentage elongation after fracture and the percentage reduction of area are very small. The tensile strength of the gray cast iron is much lower than its compressive strength, so it is a typical brittle material.

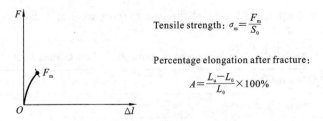

Tensile strength: $\sigma_m = \dfrac{F_m}{S_0}$

Percentage elongation after fracture:

$$A = \frac{L_u - L_0}{L_0} \times 100\%$$

**Figure 1-3    Tensile Curve for the Gray Cast Iron**

## V. Experiment Steps

(1) Measure the geometry. Each group is expected to select specimens, a low carbon

steel and a gray cast iron, and to discriminate the material by listening the sound, then wipe it clean. And then, use a vernier caliper to measure the diameter and gauge length of the specimen and record them.

(2) Stretch the specimens. According to the operating procedures, use the electronic universal testing machine to stretch the specimen, observe the phenomenon of the material in stretching until the specimen is broken, and record the load data.

(3) Measure the size of the specimen after fracture. Take off the specimen, press the fractured parts tightly, and measure the diameter and gauge length with a vernier caliper.

(4) Operate the electronic universal testing machine. The operating procedures are as follows:

① Boot sequence: turn on the main switch of the testing machine, then turn on the computer and enter into the operation window.

② Select appropriate clamps according to the shape and the size of the specimen.

③ Move the upper beam into a suitable position by the control box on the gatepost. First, install the upper clamp, click the "force reset" on the interface, then move the beam and install the lower clamp.

④ Create a new specimen in the computer and input the information of the specimen. The numbers for the specimen are the low carbon steel—XY1, the gray cast iron—XY2(X represents the batch, Y represents the group number).

⑤ Select a proper speed. Reset the displacement and time, and then perform the experiment. For the low carbon steel, the material can be elongated at the speed of 2~5 mm/min before yielding; after the material is in strengthening, it can be loaded at a speed of 10~20 mm/min. For other materials, please select the speed according to the corresponding standards.

⑥ Save the experiment data and result, generate an experiment report when the specimen is broken. And then perform the experiment analysis: experiment analysis→experiment report→ obtain tensile graph and load data.

⑦ Remove the specimen and install another one.

## Ⅵ. Recording the Experiment Data and Calculating

Please record the experiment data of the low carbon steel and the gray cast iron in Table 1-1 and Table 1-2.

**Table 1-1  Tensile Experiment Data of the Low Carbon Steel**

| Measured data | | | 1st measurement | 2nd measurement | Average |
|---|---|---|---|---|---|
| Before fracture | Section 1 diameter | $d_{01}$/mm | | | |
| | Section 2 diameter | $d_{02}$/mm | | | |
| | Section 3 diameter | $d_{03}$/mm | | | |
| | Original gauge length | $L_0$/mm | | | |

| Measured data | | | 1st measurement | 2nd measurement | Average |
|---|---|---|---|---|---|
| After fracture | Minimum cross-section diameter | $d_1$/mm | | | |
| | Final gauge length after fracture | $L_u$/mm | | | |
| | Yield load/kN | $F_{eL}$ | | | |
| | Maximum load/kN | $F_m$ | | | |

Table 1-2   Tensile Experiment Data of the Gray Cast Iron

| Measured data | | | 1st measurement | 2nd measurement | Average |
|---|---|---|---|---|---|
| Before fracture | Section 1 diameter | $d_{01}$/mm | | | |
| | Section 2 diameter | $d_{02}$/mm | | | |
| | Section 3 diameter | $d_{03}$/mm | | | |
| | Original gauge length | $L_0$/mm | | | |
| After fracture | Maximum load/kN | $F_m$ | | | |

## 1. Accuracy of Calculation

(1) The calculation accuracy of the strength indexes (lower yield strength $\sigma_{eL}$ and tensile strength $\sigma_m$) is required to be 0.5 MPa, that is, the value small than 0.25 MPa is truncated, and the value small than 0.75 MPa but equal to or bigger than 0.25 MPa is reduced to 0.5 MPa, and the value equal to or bigger than 0.75 MPa is converted to 1 MPa.

(2) The calculation accuracy of the plasticity indexes (percentage elongation after fracture $A$ and percentage reduction of area $Z$) is required to be 0.5%, that is, the value small than 0.25% is rounded off and the value small than 0.75% but equal to or bigger than 0.25% is reduced to 0.5%, and the value equal to or bigger than 0.75% is rounded to 1%.

## 2. Discussion and Analysis

(1) Fracture characteristics of the low carbon steel tensile specimen: The fracture surface of the low carbon steel specimen is like a cup cone with necking phenomenon. The middle area of the fracture is rough and showing brittle fracture (being pulled), and the outside of the fracture is smooth and is plastic deformation area with 45° shear lip (being cut).

(2) Fracture characteristics of the gray cast iron tensile specimen: The fracture surface of the gray cast iron specimen is concave, uneven and granular, the whole cross-section is about perpendicular to the axis of the specimen, no necking, and it's a typical brittle fracture.

## VII. Cautions

"Fast" loading is strictly prohibited during the experiment. The loading speed should be even and slow to prevent impact. In case of failure, stop the machine immediately and contact the teacher.

## Ⅷ. Questions

(1) Draw the tensile curves of the low carbon steel and the gray cast iron, and mark the key points.

(2) What are the similarities and differences between the mechanical properties and failure modes of the low carbon steel and the gray cast iron with normal temperature and static load?

(3) What is the practical value of measuring the mechanical properties of materials?

(4) What do you think are the factors causing the error to the test results? How can their influence be avoided or minimized?

# Experiment 2  Torsion Experiment of Metallic Materials

### I. Objectives

(1) To determine the strength indexes of metallic materials in torsion: the lower yield strength $\tau_{eL}$ and torsional strength $\tau_m$ of the low carbon steel; the torsional strength $\tau_m$ of the gray cast iron.

(2) To draw the fracture diagrams of the low carbon steel and the gray cast iron, and compare the torsion failure forms of them.

### II. Experiment Equipment and Instruments

(1) A torsion testing machine.

(2) A vernier caliper.

### III. Specimen

According to the National Standard GB/T 10128-2007 *metallic materials—torsion test at ambient temperature*, the torsion metallic specimens can be divided into circular section specimens and tubular section ones according to the product varieties, specifications and test purposes. Among them, the most commonly used is the circular section specimen, as shown in Figure 1-4. Usually, the diameter of the circular section specimen is suggested to be $d=10$ mm, the gauge length $L_0$ is 50 mm or 100 mm, and the length of the parallel part should be $L_0+2d$. If other specimens with different diameters are used, the length of the parallel part shall be the gauge length plus $2d$ too. The shape and the size of the head of the specimen should be suitable for the clamps of the torsion testing machine.

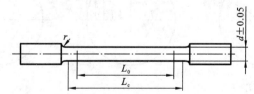

**Figure 1-4  A Specimen with a Circular Section**

Due to the maximum shear stress occurring on the surface of the specimen during the torsion experiment, the surface defects of the specimen will affect the experiment results sensitively. Therefore, the requirement of the surface roughness of the torsion specimen is higher than that of the tensile specimen. The technical processing requirements of the torsion specimen can be found in the National Standards.

## IV. Experiment Principles

### 1. To Determine the Strength of the Low carbon Steel in Torsion

Under the action of the moment of external couple, any point on the specimen is in a state of pure shear stress. With the increase of the moment of external couple, the specimen changes from elastic to yielding, then to strengthening, and broken finally, as shown in Figure 1-5. The lower yield strength of the low carbon steel is

$$\tau_{eL} = \frac{3}{4}\frac{T_{eL}}{W}$$

where $W = \pi d^3 / 16$ is the torsional cross-section coefficient of the specimen within the gauge length.

The torsional strength of the low carbon steel is

$$\tau_m = \frac{3 T_m}{4W}$$

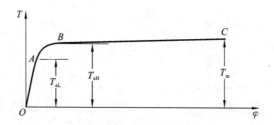

**Figure 1-5    The Torsion Curve of the Low Carbon Steel**

It can be seen from Figure 1-6 that the torque on the cross-section of the cylindrical specimen is

$$T_{eL} = \int_0^{d/2} \tau_{eL}\rho 2\pi\rho d\rho = 2\pi\tau_{eL}\int_0^{d/2} \rho^2 d\rho = \frac{\pi d^3}{12}\tau_{eL} = \frac{4}{3}W\tau_{eL}$$

$\tau_{eL}$ can then be deduced.

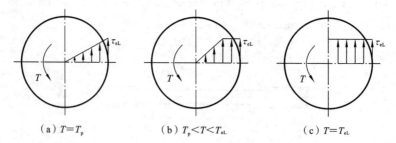

(a) $T=T_p$         (b) $T_p<T<T_{eL}$         (c) $T=T_{eL}$

**Figure 1-6    Distribution of Shear Stress in Torsion of the Low Carbon Steel Cylindrical Specimen**

According to the curve in Figure 1-5, when the moment of external couple is bigger than $T_{eL}$, the moment of external couple $T$ increases very slightly with the torsion angle $\varphi$ increasing, line $BC$ is nearly a straight line. Therefore, it can be considered that the shear stress on the cross-section is distributed as shown in Figure 1-6, excepting that the shear stress is larger than $\tau_{eL}$. According to the measured moment of external couple $T_m$ when breaking, the torsional strength can then be calculated, and the value is

$$\tau_m = \frac{3}{4} \frac{T_m}{W}$$

## 2. To Determine the Strength of the Gray Cast Iron in Torsion

For the gray cast iron specimen, we only need to measure the maximum moment of external couple $T_m$, and then the torsional strength can be calculated, the value is

$$\tau_m = \frac{T_m}{W}$$

It can be seen from the torsional fractured specimens that the fracture of the low carbon steel specimens is perpendicular to the axis, indicating that the failure is caused by the shear stress. But the fracture of the gray cast iron specimens follows a helical direction at an angle of about 45° from the axis, indicating that the damage is caused by the tensile stress.

The specimen is subject to a torsion and the material is in a state of pure shear stress, as shown in Figure 1-7. The helical surface at a 45° angle from the axis suffers the principal stresses $\sigma_1$ and $\sigma_3$. The tensile strength of the low carbon steel is greater than the shearing strength, so it is sheared to broke at a cross section, while the tensile strength of the gray cast iron is weaker than the shearing strength, so it is pulled to broke along the direction at an angle of about 45° from the axis.

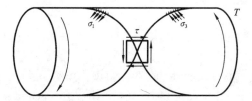

**Figure 1-7   The State of Pure Shear Stress**

## V. Experiment Steps

(1) Measure the specimen's geometry. Measure the diameters at both ends and three positions in the middle of the gauge length along the mutually perpendicular direction, and calculate the average diameter of the specimen. The minimum diameter is then used to calculate the original cross-section area $S_0$ and gauge length $L_0$.

(2) Install the specimens. First, install the moving end of the clamp, and reset the force in the computer, then install the fixed end.

(3) Computer operation. Choose a schematic and enter the specimen information.

(4) Load. Run the torsion testing machine according to the rules.

(5) Save the data and record graphics and torque when the machine stops automatically.

## Ⅵ. Recording the Experiment Data and Calculating

Record the experiment data, and determine the strength indexes of the low carbon steel and gray cast iron in torsion in Table 1-3 and Table 1-4.

**Table 1-3　Data of the Low Carbon Steel**

| Data of the low carbon steel | | 1st measurement | 2nd measurement | Average |
|---|---|---|---|---|
| Section 1 diameter | $d_{01}$/mm | | | |
| Section 2 diameter | $d_{02}$/mm | | | |
| Section 3 diameter | $d_{03}$/mm | | | |
| Yield strength | $T_{eL}$ | | | |
| Maximum torque/(N·m) | $T_m$ | | | |

**Table 1-4　Data of the Gray Cast Iron**

| Data of the gray cast iron | | 1st measurement | 2nd measurement | Average |
|---|---|---|---|---|
| Section 1 diameter | $d_{01}$/mm | | | |
| Section 2 diameter | $d_{02}$/mm | | | |
| Section 3 diameter | $d_{03}$/mm | | | |
| Maximum torque/(N·m) | $T_m$ | | | |

## Ⅶ. Questions

(1) Draw the torsion curves of the low carbon steel and the gray cast iron specimens, and mark the key points.

(2) Draw the fracture diagrams of the low carbon steel and the gray cast iron specimens.

(3) Compare the torsional failure fractures of the low carbon steel and the gray cast iron specimens, and analyse the failure reasons.

# Chapter 2

# Stress Analysis Experiments by Electrical Measurement Method

Electrical measurement method is one of the most widely used and effective methods in stress analysis. It's widely used in engineering and technical fields such as machinery, civil engineering, water conservancy, aerospace, and etc. , and it's a powerful means for verifying the basic theories of material mechanics, testing the engineering quality and carrying out scientific researches.

## Experiment 3    Basic Electrical Measurement Method and the Use of Static Resistance Strain Indicator

**Electrical resistance strain measurement** is a method which converts strain into electrical signal for measurement, referred to as electrical measurement method. The basic principle for electrical measurement method is: sticking an electrical resistance strain gauge (named as strain gauge for short) on the surface of component to be measured, when the component deform, the strain gauge will deform at the same time, the resistance strain gauge will detect corresponding change by using the electrical resistance strain measuring indicator (named as **resistance strain indicator** for short), and convert it into strain values, or output the analog electrical signal which is proportional to the strain (voltage or current), then use a recorder to record, and the measurement of the strain is realized.

The strain gauge is with light weight and small volume, and it can be used in special environment such as high (low) temperature or high pressure. The output in the measurement process is electric signal when using the electrical measurement method, which is convenient for further automation and digitalization, so the electrical measurement method is characterized by high sensitivity, and can be used for remote or wireless measurement.

### I . The Structure of Electrical Resistance Strain Gauge

The structure of the resistance strain gauge is very simple: A very thin wire with high resistivity is arranged and wound on the producing machine as shown in Figure 2-1, after that, it is bonded between two pieces of thin sheets with glue, and then the thicker lead wires are welded, and this is the early common wire-wound strain gauge. Generally, a strain gauge is composed of five parts: sensitive gates (i. e. , a metallic wire), binder, a substrate, outlet wires and covering layer. If the strain gauge is mounted on the surface of a component during test, the electrical resistance value of the wire will change when it deforms together with the component.

Commonly used strain gauges include the wire-wound strain gauge (Figure 2-1), the short-wire strain gauge and the foil strain gauge (Figure 2-2). They all belong to the uniaxial strain gauges, that is, there is only one sensitive gate on a base, which is used to measure the strain along the gate axis. As shown in Figure 2-3, several sensitive gates are arranged at a certain angle on the same base to measure the strains along the axis of several sensitive gates at the same point, so it is called **multi-axis strain gauge**, commonly known as **strain rosette**. A strain rosette is mainly used to measure the principal strain and direction of a point in plane stress state.

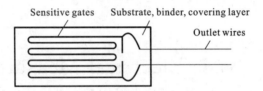

Figure 2-1　Wire-wound Strain Gauge

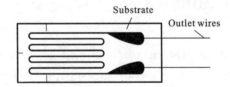

Figure 2-2　Foil Strain Gauge

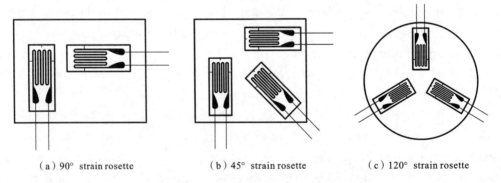

(a) 90° strain rosette　　　(b) 45° strain rosette　　　(c) 120° strain rosette

Figure 2-3　Different Strain Rosettes

## Ⅱ. Sensitivity Coefficient of Resistance Strain Gauge

When the strain gauge is used for strain measurement, a certain voltage is applied to the wire in the strain gauge. In order to prevent excessive current, heating and fusing, the wire must have a certain length to obtain a large initial resistance value. However, when measuring the strain of components, it is required to shorten the length of wire as much as possible to measure the real strain of "a point". Therefore, the wire in the strain gauge is generally made into a grid as shown in Figure 2-1, which is called the sensitive gate. The phenomenon that the resistance value of the wire in the strain gauge pasted on the component changes with the deformation of the component is called resistance strain phenomenon. When the strain gauge is mounted on the surface of a specimen with an unidirectional stress and the grid axis direction of the sensitive gate is consistent with the stress direction, within a certain deformation range, the change rate of the resistance of the strain gauge $\Delta R/R_0$ is proportional to the strain $\varepsilon$ in the axial direction of the sensitive gate, i. e.

$$\frac{\Delta R}{R_0} = K\varepsilon$$

where $R_0$ is the original resistance value of a strain gauge; $\Delta R$ is the change of resistance value of a strain gauge; $K$ is called the sensitivity coefficient of a strain gauge.

The sensitivity coefficient of a strain gauge is generally determined by the manufacturer through experiments. This step is called the calibration of strain gauge. In practical applications, the strain gauges with different sensitivity coefficients should be selected according to the needs.

## II. Measuring Circuit of the Resistance Strain Gauge

When strain gauges are used to measure the strains, the small changes in resistance must be measured in an appropriate manner. Therefore, the strain gauge is usually connected to a certain circuit so that the change of its resistance value can change the circuit in a certain way and make the circuit output a signal that can indicate the change of the resistance value of the strain gauge. Then, the strains can be obtained by processing the electrical signal only. The input circuit of the resistance strain gauge used in the electrical measurement method is called the strain bridge, which is a four-arm bridge with the strain gauge as a part or the whole of the bridge arms. It can transform the tiny change of strain gauge resistance value into the change of output voltage. The DC voltage bridge is taken as an example to illustrate in this book.

### 1. Output Voltage of the Bridge

The bridge circuit of the resistance strain gauge is shown in Figure 2-4. It takes the strain gauge or resistance element as the bridge arm, e. g. , taking $R_1$ as a strain gauge, or $R_1$ and $R_2$ as strain gauges, or $R_1$-$R_4$ as strain gauges. Ports $A$, $C$ and $B$, $D$ are the input and output ports of the bridge, respectively.

According to the principle of electrotechnics, it can be deduced that when the input ports have a voltage $U_I$, the output

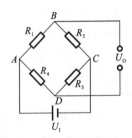

**Figure 2-4　Bridge Circuit**

voltage of the bridge is

$$U_O = \frac{R_1 R_3 - R_2 R_4}{(R_1 + R_2)(R_3 + R_4)} U_I$$

When $U_O = 0$, the bridge is in equilibrium. Therefore, the equilibrium condition of the bridge is $R_1 R_3 = R_2 R_4$. When the resistance value of each bridge arm in the balanced bridge varies with $\Delta R_1, \Delta R_2, \Delta R_3$ and $\Delta R_4$ respectively, the output voltage of the bridge can be approximately calculated as

$$U_O \approx \frac{U_I}{4} \left( \frac{\Delta R_1}{R_1} - \frac{\Delta R_2}{R_2} + \frac{\Delta R_3}{R_3} - \frac{\Delta R_4}{R_4} \right)$$

It can be seen that the strain bridge has an important property: the output voltage of the strain bridge is proportional to the difference of the resistance change rate of the two opposite bridge arms and the sum of the resistance change rate of the two opposite bridge arms. For a balanced bridge, if the resistance change rate of the two adjacent bridge arms is equal and increasing or decreasing simultaneously, or the resistance change rate of the two opposite bridge arms is equal but increasing or decreasing mutually-exclusively, the equilibrium state of the bridge will not change, that is, it will remain $U_O = 0$. If all four arms of the bridge are connected with the same strain gauge, then,

$$U_O = \frac{K U_I}{4} (\varepsilon_1 - \varepsilon_2 + \varepsilon_3 - \varepsilon_4)$$

where $\varepsilon_1 \text{-} \varepsilon_4$ are the strain values of the strain gauges connected to the four bridge arms of the electric bridge.

## 2. Compensation of Temperature Effect

The specimen mounted with strain gauges is always in a certain temperature field. If the linear expansion coefficients of the sensitive gate material and the component are not equal, when the temperature changes, the percentage elongation or shorten of the sensitive gate and the component won't be equal, and then the sensitive gate will be stretched (compressed) additionally, which will lead to the change of the resistance of the sensitive gate, this phenomenon is called the temperature effect. The change rate of sensitive gate resistance with temperature can be approximately regarded as a proportional one. The change of temperature has a great influence on the output voltage of the bridge. In a serious case, dozens of microstrains ($\mu\varepsilon$) can be generated in the resistance strain gauge with only increasing 1 ℃, so we need to take actions to exclude the effect of temperature. The method that excludes the effect of temperature is called temperature compensation.

According to the property of the bridge, temperature compensation is not difficult. We can use a strain gauge as a temperature compensation gauge, mounted it on a specimen with the same material as a specimen to be measured but without stress during test. This specimen and the specimen to be measured are placed together during test so that they are in the same temperature field. The strain gauge mounted on the specimen to be measured is called the working gauge. When connecting to the bridge, the working gauge and the temperature compensation gauge are placed at two adjacent bridge arms, as shown in Figure 2-5. Since the temperature of the working gauge and the temperature compensation gauge is always the same, the change of the resistance value caused by the temperature change is also the same, and they are in the two adjacent arms of the bridge, the influence of the temperature effect is then eliminated.

The resistance value, sensitivity coefficient and resistance temperature coefficient of the working gauge and the temperature compensation gauge should be the same, and they should be mounted on the specimen to be measured and a specimen with no force respectively, so as to ensure that the change of resistance value of the strain gauge caused by the change of temperature is the same.

### 3. Strain Gauge Layout and Its Connection in the Bridge

The strain gauge reflects the tensile or compressive strain at a point on the surface of the component. In some cases, the strain may be related to a variety of internal forces (such as axial forces and bending moments). Thus, the strain corresponding to some internal forces is simply measured and the strain corresponding to other internal forces should be removed from the total strains. Obviously, the strain gauge itself cannot distinguish various strain components. However, as long as the position and direction of the pasted strain gauge is reasonably selected and the strain gauge is reasonably connected to the bridge, the specified strain of a certain point can be measured from the complex composite strains by using the properties of the bridge.

Usually, there are three types for a strain gauge connected into a bridge.

(1) **Half-bridge single-arm connection method**   It is also called 1/4-bridge connection method. As shown in Figure 2-5, a working gauge and a temperature compensation gauge are connected to two adjacent bridge arms respectively, and the other two bridge arms are connected with resistors. If the strain of the working gauge is $\varepsilon_1$, then the output voltage of the bridge is

$$U_O = \frac{K U_I}{4}\varepsilon_1$$

and the resistance strain indicator shows $\varepsilon = \varepsilon_1$.

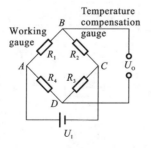

Figure 2-5   Half-bridge Single-arm Connection Method

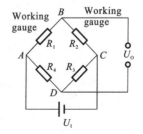

Figure 2-6   Half-bridge Double-arm Connection Method

(2) **Half-bridge double-arm connection method**   As shown in Figure 2-6, two working gauges are connected to two adjacent bridge arms, and the other two bridge arms are con-

nected with resistors. The two working gauges are temperature compensation gauges for each other at the same time. If the strains of the working gauges are $\varepsilon_1$ and $\varepsilon_2$, $\varepsilon_1 = -\varepsilon_2$, the output voltage of the bridge is

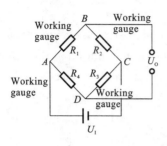

**Figure 2-7 Full-bridge Connection Method**

$$U_O = \frac{KU_I}{4}(\varepsilon_1 - \varepsilon_2) = \frac{KU_I}{2}\varepsilon_1$$

and the resistance strain indicator shows $\varepsilon = 2\varepsilon_1$.

(3) **Full-bridge connection method**　　As shown in Figure 2-7, four arms of the bridge are connected with the working gauges, and suppose that the strain of each working gauge is $\varepsilon_1$, $\varepsilon_2$, $\varepsilon_3$ and $\varepsilon_4$ respectively, and $\varepsilon_1 = -\varepsilon_2 = \varepsilon_3 = -\varepsilon_4$, thus, the output voltage of the bridge is

$$U_O = \frac{KU_I}{4}(\varepsilon_1 - \varepsilon_2 + \varepsilon_3 - \varepsilon_4) = KU_I\varepsilon_1$$

and the resistance strain indicator shows $\varepsilon = 4\varepsilon_1$.

### Ⅳ. Static Resistance Strain Indicator

A static resistance strain indicator is a kind of resistance strain instrument which is specially designed to measure the resistance strain without changing or changing very slowly with time. Its function is to enlarge the output voltage of the strain bridge, to display the value of strain in scale or digital form in the display screen, or to supply the recorder with the analogical electric signal of the strain change.

### 1. Working Principle of the Static Resistance Strain Indicator

There are many kinds of static resistance strain indicators. In this book, the CML-1H static resistance strain indicator is taken as an example to introduce the working principles.

The block diagram of the working principle of CML-1H static resistance strain indicator is shown in Figure 2-8. The instrument has a built-in microprocessor to realize functions such as acquisition, processing, displaying and communication. The 1/4-bridge, half-bridge or full-bridge can be chosen to measure. When measuring the strain, the strain gauge mount-

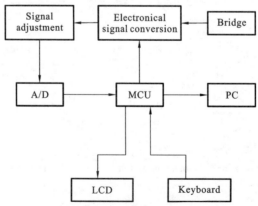

**Figure 2-8　Block Diagram of the Working Principle of CML-1H Static Resistance Strain Indicator**

ed on the specimen is connected to the bridge, and the bridge is adjusted and balanced. When the specimen is subjected to external forces and deformation, the resistance value of the strain gauge changes accordingly, thus, the balance of the bridge is damaged and the output voltage is generated. The value of strain is shown in the display screen. The static resistance strain indicator can measure 16 points' strains at the same time. It can also be connected to a computer through a connecting interface, then processing the measured data with a computer.

## 2. Wiring Methods of the Bridge for CML-1H Static Resistance Strain Indicator

The wiring methods of the bridge for CML-1H static resistance strain indicator are shown in Figures 2-9,2-10,2-11 and 2-12.

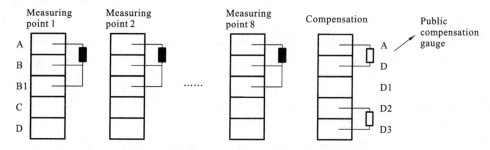

**Figure 2-9   Wiring Method for a 1/4-bridge**

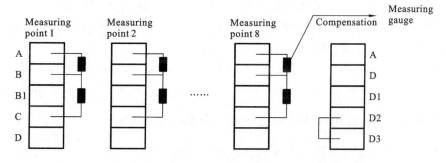

**Figure 2-10   Wiring Method for a Half-bridge**

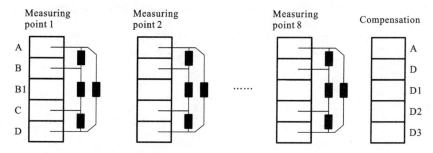

**Figure 2-11   Wiring Method for a Full-bridge**

## 3. Application of the CML-1H Static Resistance Strain Indicator

(1) Turn on the indicator. Turn the AC 220 V power switch to "on". At the same time, the 9 sets of digital tubes will light up, and the displaying decreases from 5 to 0. After

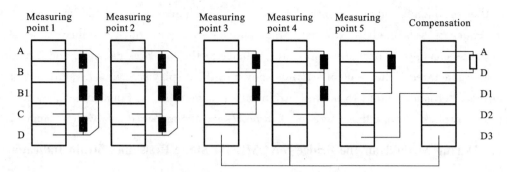

Figure 2-12　Combined Wiring Method

the self-test, the display part of the machine will flicker. Directly click "ok" button and pre-heat it for 30 minutes.

(2) Connect wires while preheating. Select the required bridge according to the experiment requirements, and connect the strain gauge to the terminals of the resistance strain indicator.

(3) Set parameter.

(4) Reset the strain.

(5) Load and record the data. In the left part of the digital panel of the resistance strain indicator, the 1st and 2nd numbers indicate the channel number, the 3rd number is the positive and negative sign, and the 4th-8th numbers are the value of the strain.

(6) When finishing load, check the data and give it to the teacher to examine. If approved, proceed to the next step.

(7) When the experiment is finished, unload, turn off the power, uninstall, dismantle the connection lines, sort out the instrument and tidy the desk at the end of the experiment.

## Ⅴ. Wiring Experiment of the Resistance Strain Indicator

Taking the normal stress in bending test device for a beam with uniform cross-section as an example, practise the half-bridge single-arm connection method (1/4-bridge connection method), half-bridge double-arm connection method and full-bridge connection method.

# Experiment 4　Measurement Experiment of Bending Normal Stress of Rectangular Beam (1/4-bridge)

## I . Objectives

(1) To be familiar with the basic principle of electrical measurement method and the use of a static resistance strain indicator.

(2) To measure the normal stress distribution on the cross-section of a rectangular beam in pure bending and in transverse bending.

## II . Experiment Equipment and Instruments

(1) A WYS-1 laboratory bench of mechanics of materials.

(2) A static resistance strain indicator.

## III . Experiment Principles

The experimental rig is shown in Figure 2-13. The rectangular beam is made of low carbon steel. 8 strain gauges are mounted on the lines with different heights parallel to the axis as working gauges on both sides of the beam, and the beam is subjected to pure bending and transverse bending deformation, two strain gauges are mounted on the steel sheet of the same material as the beam as temperature compensation gauges.

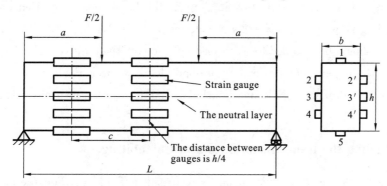

**Figure 2-13　A Schematic of the Normal Stress for a Beam with Uniform Cross-section Area in Bending**

The 16 working gauges and temperature compensation gauges are connected to the channel on the static resistance strain indicator panel in the form of 1/4-bridge respectively. When the beam is subjected to bending deformation, the resistance value of the working gauges will change as the deformation of the beam, and the strain value $\varepsilon_{real}$ of each corresponding position can be measured by the static resistance strain indicator. The distribution of the strain along the height of the beam is shown in Figure 2-14.

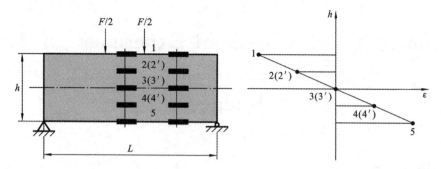

**Figure 2-14    Distribution of the Strain along the Height of the Beam**

The method of equivalent step loading is adopted. According to Hooke's law, the corresponding stress value of each point can be calculated by

$$\sigma_{real} = E\varepsilon_{real}$$

where $E$ is the elastic modulus of the material of the beam.

When the beam is bending, the theoretical calculation formula of the normal stress on the cross-section is

$$\sigma_{theory} = \frac{My}{I_z}$$

where $I_z = bh^3/12$ is the moment of inertia of the cross-section of the beam to the neutral axis, $y$ is the distance between the neutral axis and the desired stress point. For pure bending deformation, $M_{pure} = Fa/2$, and for transverse bending deformation, $M_{transverse} = F(a-c/2)/2$.

## IV. Experiment Steps

(1) Wiring. Use the 1/4-bridge. Please refer to the manual for detailed operation steps.

(2) Loading. The steady and slow loading of the handwheel is carried out with equal loads, and loading step by step ($F_1 = 1$ kN, increment 0.5 kN, $F_{max} = 3$ kN). The strain values at each point should be recorded for each load (please note the positive and negative signs).

(3) When the experiment is finished, unload slowly, check the data with the teacher. Turn off the power, dismantle the connection lines, and arrange the instrument back to its position.

## V. Recording the Experiment Data and Calculating

The geometry of the rectangular beam is recorded in Table 2-1, and the data of the rectangular beam in the bending experiment and transverse bending experiment is recorded in Table 2-2 and Table 2-3 respectively.

**Table 2-1    Geometry of the Rectangular Beam**

| $a$/mm | $b$/mm | $c$/mm | $L$/mm | $h$/mm | $E$/GPa | $K$ | $R_0$ |
|--------|--------|--------|--------|--------|---------|-----|-------|
|        |        |        |        |        |         |     |       |

**Table 2-2    Data of the Rectangular Beam in the Bending Experiment**

| Load/kN | Strains($\mu\varepsilon$) | | | | | | | |
|---|---|---|---|---|---|---|---|---|
| | $y_1 = 0$ | $y_2 = 0$ | $y_3 = h/4$ | $y_4 = h/4$ | $y_5 = -h/4$ | $y_6 = -h/4$ | $y_7 = h/2$ | $y_8 = -h/2$ |
| 1 | | | | | | | | |
| 1.5 | | | | | | | | |
| 2 | | | | | | | | |
| 2.5 | | | | | | | | |
| 3 | | | | | | | | |
| $\overline{\varepsilon_i} =$ | $\overline{\varepsilon_1} =$ | $\overline{\varepsilon_2} =$ | $\overline{\varepsilon_3} =$ | $\overline{\varepsilon_4} =$ | $\overline{\varepsilon_5} =$ | $\overline{\varepsilon_6} =$ | $\overline{\varepsilon_7} =$ | $\overline{\varepsilon_8} =$ |
| $\sigma_{real} = E\overline{\varepsilon}$ | | | | | | | | |
| $\sigma_{theory} = \dfrac{\overline{M}y}{I_z}$ | | | | | | | | |

**Table 2-3    Data of the Rectangular Beam in the Transverse Bending Experiment**

| Load/kN | Strains($\mu\varepsilon$) | | | | | | | |
|---|---|---|---|---|---|---|---|---|
| | $y_1 = 0$ | $y_2 = 0$ | $y_3 = h/4$ | $y_4 = h/4$ | $y_5 = -h/4$ | $y_6 = -h/4$ | $y_7 = h/2$ | $y_8 = -h/2$ |
| 1 | | | | | | | | |
| 1.5 | | | | | | | | |
| 2 | | | | | | | | |
| 2.5 | | | | | | | | |
| 3 | | | | | | | | |
| $\overline{\varepsilon_i} =$ | $\overline{\varepsilon_1} =$ | $\overline{\varepsilon_2} =$ | $\overline{\varepsilon_3} =$ | $\overline{\varepsilon_4} =$ | $\overline{\varepsilon_5} =$ | $\overline{\varepsilon_6} =$ | $\overline{\varepsilon_7} =$ | $\overline{\varepsilon_8} =$ |
| $\sigma_{real} = E\overline{\varepsilon}$ | | | | | | | | |
| $\sigma_{theory} = \dfrac{\overline{M}y}{I_z}$ | | | | | | | | |

## Ⅵ. Cautions

(1) Check the position of the rectangular beam and try to make the center line of the pressure head pass through the longitudinal axisymmetric plane of the beam, to ensure that the middle of the rectangular beam is subjected to pure bending deformation.

(2) It should be confirmed that no load is loaded on the beam during the balance operation of the static resistance strain indicator, and the "reading" of the force should be "0" (If it is not zero, record the initial value).

(3) The handwheel should be rotated smoothly during loading, never too fast. When the force value is basically stable at the specified value, stop rotating the handwheel and measure the strains.

(4) The maximum load is 3000 N. Overload is strictly prohibited to avoid damage to the

force sensor.

(5) The wiring shall not be loose during the measurement, and the wires with the strain gauge shall not be moved or contacted in a load measurement cycle, so as to ensure the stability and reliability of the strain value measured.

## Ⅶ. Data Processing

(1) Incremental strain calculation. According to the recorded strain values at each point, calculate the 4 strain increments at each point. According to Hooke's law, 4 strain increments measured at the same point should be the same. The reasons should be found if the strain increments at the same point are different.

(2) Calculation of measured stress at each point. Calculate the average strain increment at each point. For the case that the measuring points located on the front and back sides of the beam, the average value of the strain increments should be calculated with the averages of strain increments on the front and back sides, and then to obtain the average value of the strain increments when the load increment is 0. 5 kN. By Hooke's law $\sigma = E\varepsilon$, the stress can be calculated. The minimum strain reading for a strain indicator is "1 $\mu\varepsilon$", meaning the strain value is "$1 \times 10^{-6}$". So, the minimum value of the measured strain increment is also 1 $\mu\varepsilon$.

(3) Error calculation and analysis. The error of the neutral layer is an absolute one, and the error of other points is a relative one.

## Ⅷ. Questions

(1) Why should the temperature compensation gauge be mounted to the same material as the specimen's?

(2) Is the bending normal stress affected by the elastic modulus of the material? Why?

# Experiment 5   Measurement Experiment of Bending Normal Stress of Rectangular Beam ( Half - bridge and Full - bridge )

### Ⅰ. Objectives

(1) To be familiar with the basic principles of electrical measurement method, learn the connection methods of half-bridge and full-bridge, and master the methods of multi-point static strain measurement further.

(2) To measure the normal stress for pure bending and transverse bending on the cross-section of a uniform beam, and compare with the results obtained by the 1/4-bridge method.

### Ⅱ. Experiment Equipment and Instruments

(1) A WYS -1 laboratory bench of mechanics of materials.

(2) A static resistance strain indicator.

(3) A single beam with rectangular cross-section.

### Ⅲ. Experiment Principles

As shown in Figure 2-15, the load $F$ on the beam is divided into two forces of equal magnitude $F/2$ acting on the rectangular steel beam. A pure bending deformation is formed in the middle of the beam, and the bending moment is $M=Fa/2$.

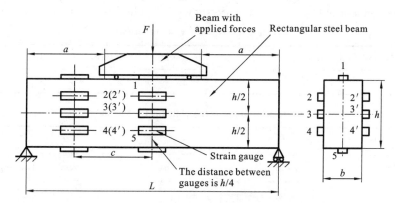

**Figure 2-15   Strain Gauges Layout on a Rectangular Beam**

Based on the neutral layer of the beam, 8 strain gauges parallel to the beam axis are mounted at every $h/4$ height on the upper and lower surfaces, and on the front and back sides for both middle position(pure bending) and outside position(transverse bending) of the loading steel, thus, there are 16 gauges in all for two positions.

The center of each strain gauge sensitive gate is the measuring point of the experiment.

According to the measured strain values of each measuring point, the measured stress values are calculated by Hooke's law ($\sigma = E\varepsilon$), and the distribution law of the normal stress in cross-section along the beam height can be obtained, and compared with the theoretical stress calculated by bending normal stress formula $\sigma = \dfrac{My}{I_z}$, where $I_z = bh^3/12$ is the inertial moment of the cross-section of the beam, $y$ is the distance between the element and the neutral axis. For pure bending deformation, $M_{pure} = Fa/2$, and for transverse bending deformation, $M_{transverse} = F(a - c/2)/2$.

## Ⅳ. Experiment Steps

(1) Wiring. Use the half-bridge method, please refer to the instructions of the static resistance strain indicator for details of operation steps.

(2) Loading. Load steadily and slowly with rotating the handwheel at a constant speed step by step ($F_1 = 1$ kN, increment $0.5$ kN, $F_{max} = 3$ kN). Strain values at each point should be recorded for each load (pay attention to the positive and negative signs).

(3) When the half-bridge experiment is finished, unload slowly, submit the data to the teacher, turn off the power and remove the connection.

(4) Wiring again. Perform the full-bridge method and repeat the above steps.

(5) When the experiment is finished, unload slowly, turn off the power and remove the connection. Tidy up the experiment table.

## Ⅴ. Recording the Experiment Data and Calculating

The geometry of the beam, the elastic modulus $E$, the sensitivity coefficient of the strain gauge $K$ and the resistance of the strain gauge $R_0$ are shown in Table 2-4. Data of the half-bridge method and full-bridge method are shown in Table 2-5 and Table 2-6 respectively.

**Table 2-4  Geometry of the Beam**

| $a$/mm | $b$/mm | $c$/mm | $L$/mm | $h$/mm | $E$/GPa | $K$ | $R_0$ |
|--------|--------|--------|--------|--------|---------|-----|-------|
|        |        |        |        |        |         |     |       |

**Table 2-5  Data of the Half-bridge Method**

| Load/kN | Strain ($\mu\varepsilon$) | | | | | |
|---------|---------|---------|---------|---------|---------|---------|
|         | Pure bending | | | Transverse bending | | |
|         | $y_1 =$ | $y_2 =$ | $y_3 =$ | $y_4 =$ | $y_5 =$ | $y_6 =$ |
| 1       |         |         |         |         |         |         |
| 1.5     |         |         |         |         |         |         |
| 2       |         |         |         |         |         |         |
| 2.5     |         |         |         |         |         |         |
| 3       |         |         |         |         |         |         |

**Table 2-6   Data of the Full-bridge Method**

| Load/kN | Strain($\mu\varepsilon$) | |
|---|---|---|
| | Pure bending | Transverse bending |
| | $y_1 =$ | $y_2 =$ |
| 1 | | |
| 1.5 | | |
| 2 | | |
| 2.5 | | |
| 3 | | |

## Ⅵ. Cautions

(1) When the strain indicator is in balance operation, it shall be confirmed that there is no load on the beam, and the force reading is "0" (record the initial value if it is not zero).

(2) The positions of the loaded beam and the rectangular steel beam should be checked during loading, make sure that the center line of the force passes through the longitudinal axisymmetric plane of the beam as close as possible, so as to ensure that the middle of the rectangular beam is subjected to a pure bending deformation.

(3) The handwheel should be rotated steadily during loading, never too fast. When the force value is basically stable at the specified value, stop rotating the handwheel and measure the strains.

(4) The maximum loading is 3000 N, and overload is strictly prohibited to avoid damage to the force sensor.

(5) Make sure that the wiring is not loose during measurement, and do not move or touch the wires of strain gauge in a load measurement cycle to ensure the stability and reliability of strain value measured.

## Ⅶ. Questions

(1) Analyze the similarities and differences between the half-bridge, full-bridge and 1/4-bridge methods in this experiment.

(2) What may be the reasons that the measured strain in the neutral layer is not zero?

(3) When the load is a constant, what is the distribution law of the strain along the height of the beam? What is the theoretical basis? Please analyze it with a theoretical formula.

(4) At the same measuring point, how does the strain change with the increase of load? What is the theoretical basis? Please analyze it with a theoretical formula.

# Experiment 6   Determination of Principal Stress Direction in Bending and Torsion Combination

## Ⅰ. Objectives

(1) To determine the magnitude and direction of the principal stress at a point in the plane stress state with resistance strain rosettes.

(2) To understand the application of strain analysis theory in plane stress state in experiments.

(3) To be further familiar with the electronical bridges for the static resistance strain indicator and static multi-point strain measurement method.

## Ⅱ. Experiment Equipment and Instruments

(1) A bending-torsion combination experiment device.

(2) A static resistance strain indicator.

## Ⅲ. Experiment Principles

### 1. Experiment Equipment and Strain Gauge Layout

Figure 2-16 is a schematic diagram of the bending-torsion combination device, and Figure 2-17 is the strain rosettes layout diagram of the bending-torsion combination device. Two 45° strain rosettes are pasted at the upper and lower measuring points $A$ and $B$ on the cross-section of the thin-walled cylinder with length $L$, and the twisting arm length is $a$. With the action of gravity of the weight, the thin-walled cylinder suffers combined deformation of bending and torsion.

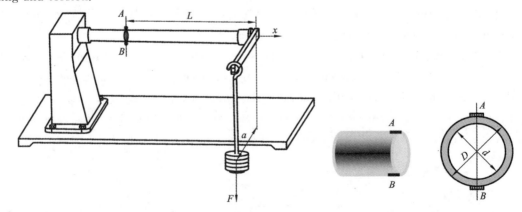

**Figure 2-16   A Schematic Diagram of the Bending-torsion Combination Device**

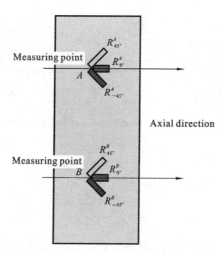

**Figure 2-17   Strain Rosettes Layout Diagram of the Bending-torsion Combination Device**

## 2. Force Analysis of Bending-torsion Combination Device

The force and force analysis diagrams of the bending-torsion combination device are shown in Figure 2-18 and Figure 2-19.

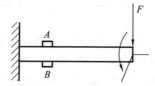

**Figure 2-18   Force Diagram of the Bending-torsion Combination Device**

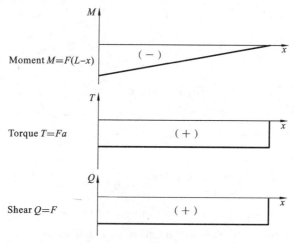

**Figure 2-19   The Force Analysis Diagram of the Bending-torsion Combination Device**

## 3. Stress Analysis for Points $A$ and $B$

Points $A$ and $B$ are the upper and lower strain rosettes sticking positions on the thin-wall cylinders, respectively. The stresses for points $A$ and $B$ are shown in Figure 2-20.

For points $A$ and $B$, the shear stress due to the torsion is

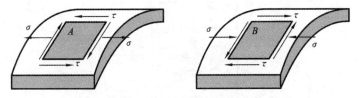

**Figure 2-20    The Stress States of Points $A$ and $B$**

$$\tau = \frac{T_a}{W_T} = \frac{Fa}{W_T}$$

The normal stress due to the bending moment is

$$\sigma = \frac{M}{W_z} = \frac{FL}{W_z}$$

It can be deduced from the above content that the theoretical principal stresses of points $A$ and $B$ are

$$\sigma_1(\sigma_3) = \sigma/2 \pm \sqrt{(\sigma/2)^2 + \tau^2}$$

And the theoretical principal direction of them is

$$\alpha = \arctan(-2\tau/\sigma)/2$$

## 4. Strain Analysis Principle of the Plane Stress State

According to the strain analysis principle of the plane stress state, the magnitude and direction of principal stress can be determined by using three different linear strains. The relationship between principal stress and principal strain of plane stress state is determined by the generalized Hooke's law:

$$\sigma_1 = \frac{E}{1-\mu^2}(\varepsilon_1 + \mu\varepsilon_3) \qquad (1)$$

$$\sigma_3 = \frac{E}{1-\mu^2}(\varepsilon_3 + \mu\varepsilon_1) \qquad (2)$$

However, the direction of the principal strain is unknown in a general case of the plane stress state, so the principal strain cannot be directly measured by the strain gauge. According to the strain analysis theory of the plane stress state, the following relationship can be found:

$$\varepsilon_a = \frac{\varepsilon_x + \varepsilon_y}{2} + \frac{\varepsilon_x - \varepsilon_y}{2}\cos2\alpha - \frac{1}{2}\gamma_{xy}\sin2\alpha \qquad (3)$$

$\varepsilon_a$ varies as angle $\alpha$, $\varepsilon_a$ will reach its maximum at the two perpendicular principal directions, those are the two principal strains. The magnitude and the direction of the two strains are

$$\varepsilon_{1,3} = \frac{\varepsilon_x + \varepsilon_y}{2} \pm \frac{1}{2}\sqrt{(\varepsilon_x - \varepsilon_y)^2 + \gamma_{xy}^2} \qquad (4)$$

$$\tan2\alpha_0 = -\frac{\gamma_{xy}}{\varepsilon_x - \varepsilon_y} \qquad (5)$$

Due to the shear strain $\gamma_{xy}$ cannot be measured by strain gauge, so one can choose any three angles, measure three line strains in three directions respectively, and substitute the strains into the three separate equations of Equation (3), solve $\varepsilon_x$, $\varepsilon_y$ and $\gamma_{xy}$ respectively.

Then, substitute $\varepsilon_x$, $\varepsilon_y$ and $\gamma_{xy}$ into Equations (4) and (5), the magnitudes and directions of the principal strains $\varepsilon_1$ and $\varepsilon_3$ then can be deduced, finally the magnitudes and directions of the principal stresses can be deduced by Equations (1) and (2), and the directions of the principal stresses are in accordance with the directions of the principal strains.

The following simple formulas can be obtained with the above equations and 45° strain rosettes.

Principal direction:

$$\tan 2\alpha_0 = \frac{\varepsilon_{45°} - \varepsilon_{-45°}}{2\varepsilon_{0°} - \varepsilon_{-45°} - \varepsilon_{45°}} \tag{6}$$

Principal strain:

$$\varepsilon_{1,3} = \frac{\varepsilon_{45°} + \varepsilon_{-45°}}{2} \pm \frac{\sqrt{2}}{2} \sqrt{(\varepsilon_{-45°} - \varepsilon_{0°})^2 + (\varepsilon_{45°} - \varepsilon_{0°})^2} \tag{7}$$

Principal stress:

$$\sigma_1 = \frac{E}{1-\mu^2}(\varepsilon_1 + \mu\varepsilon_3), \quad \sigma_3 = \frac{E}{1-\mu^2}(\varepsilon_3 + \mu\varepsilon_1) \tag{8}$$

## Ⅳ. Experiment Steps

The bridge circuit of strain indicator is still connected with 1/4-bridge circuit. Normal manual loading is adopted, each weight has a gravity of 9.8 N, and each group has 5 equal weights.

## Ⅴ. Recording the Experiment Data and Calculating

Geometries of the thin-wall cylinder and the parameters of strain rosette are shown in Table 2-7 and strains measured for points $A$ and $B$ are shown in Table 2-8.

**Table 2-7   Geometries of the Thin-wall Cylinder and the Parameters of Strain Rosette**

| Outer diameter of the thin-wall cylinder $D$/mm | Inner diameter of the thin-wall cylinder $d$/mm | $L$/mm | $a$/mm | Poisson's ratio $\mu$ | Elastic modulus $E$/GPa | Sensitivity coefficient $K$ | Resistance $R_0$ |
|---|---|---|---|---|---|---|---|
| | | | | | | | |

**Table 2-8   Strains Measured for Points $A$ and $B$**

| Load/N | Strain ($\mu\varepsilon$) | | | | | |
|---|---|---|---|---|---|---|
| | Upper point $A$ | | | Lower point $B$ | | |
| | $\varepsilon_{-45°}$ | $\varepsilon_{0°}$ | $\varepsilon_{45°}$ | $\varepsilon_{-45°}$ | $\varepsilon_{0°}$ | $\varepsilon_{45°}$ |
| 9.8 | | | | | | |
| $2\times9.8$ | | | | | | |
| $3\times9.8$ | | | | | | |
| $4\times9.8$ | | | | | | |
| $5\times9.8$ | | | | | | |
| $\overline{\varepsilon_i}=$ | $\overline{\varepsilon_{-45°}}=$ | $\overline{\varepsilon_{0°}}=$ | $\overline{\varepsilon_{45°}}=$ | $\overline{\varepsilon_{-45°}}=$ | $\overline{\varepsilon_{0°}}=$ | $\overline{\varepsilon_{45°}}=$ |

## Ⅵ. Questions

(1) How to determine the positions of the dangerous points on the cross-section and the stress state of the point when the thin-wall cylinder deformation is combined with bending and torsion?

(2) In the plane stress state, how to determine the magnitude and direction of the principal stress by measuring the linear strains in three different directions of a point?

# Experiment 7   Strain Gauges Mounting

## Ⅰ. Objectives

(1) To master the mounting technique of a strain gauge.

(2) To learn the general methods for inspecting mounting quality.

## Ⅱ. Experiment Equipment and Instruments

(1) A static resistance strain indicator.

(2) A beam with equal strength.

(3) A digital multimeter.

(4) Strain gauges, 502 glue, connecting wires and terminals.

(5) Other tools and materials, such as cleaning materials including sand paper, acetone, anhydrous alcohol, cotton wool, and tools including electric soldering iron, tweezers and rulers.

## Ⅲ. Strain Gauge Sticking Process

In the electrical stress analysis, the strain on the component surface is transmitted to the strain gauge through the adhesive layer. The reliability of the measured data is largely dependent on the mounting quality of the strain gauge. This demands that the adhesive layer is thin and uniform with no bubbles, and solidified completely, neither creeping nor degumming. The strain gauges should be mounted manually. To achieve accurate location and excellent quality, it totally depends on repeated practice to accumulate experience. The mounting process of the strain gauge includes the following processes.

### 1. Strain Gauge Screening

The wire grid or gauge grid of the strain gauge should be arranged in a neat way without bending, rust or spots, and the base should not be damaged. For the strain gauges after screening, the electrical resistance values shall be measured one by one with a digital multimeter, and the difference of resistance values shall not exceed 0.5 Ω.

### 2. Surface Treatment for the Specimen

In order to make the strain gauge stick firmly, the paint layer of the part where the strain gauge is stuck to on the specimen should be scraped off, the rust spot should be polished, and the oil stains should be removed. The surface roughness should reach up to $Ra$ 20 ~25. The treated surface area for the gauge shall be greater than three times of the base area of the strain gauge. If the surface is too smooth, polish out lines of 45° with the axis of the strain gauges by using a fine sand paper and mark a line along the mounting direction with a scriber after polishing smoothly.

Clean the polished part of the specimen before mounting the strain gauge with cotton wool or gauze dipped in acetone or anhydrous alcohol until there are no stains on the cotton wool. After acetone or anhydrous alcohol volatilizes, the surface is dry, then mount the strain gauge.

### 3. Strain Gauge Mounting

The adhesives for the normal temperature strain gauge include 502 (or 501) quick-drying glue, epoxy resin adhesive, phenolic resin adhesive, and etc. In cold or wet environment, you'd better use hot air of the hair dryer to make the mounted part of the specimen be heated to 30~40 ℃ before mounting the strain gauge. When mounting the strain gauge, apply a thin layer of adhesive to the mounted surface first, the outlet wire of the strain gauge should be pinched with fingers (or pinched with forceps), and apply the glue to the substrate which should be placed on the specimen immediately, and the base line of the strain gauge should be aligned with the mounting line marked on the specimen, and then roll the transparent PVC film (or the cellophane) with your thumb in one direction along the axis of the strain gauge. Press the strain gauge from light to heavy, squeeze out bubbles and excess glue to ensure that the adhesive layer is as thin and uniform as possible and avoid the strain gauge sliding or rotating. Press it for about half a minute to make the strain gauge stick firmly. After an appropriate drying time, gently peel off the PVC film, and observe the strain gauge. If there are bubbles in the sensitive gate, the strain gauge should be removed and cleaned and re-mounted. If the sensitive gate parts are glued, only the edge of the substrate is upturned, then just stick these local areas.

After the strain gauge is mounted, it can only be used after the adhesive has been fully cured. Different types of adhesive have different curing requirements, 502 glue can cure naturally. Before the adhesive is solidified, lift the strain gauge's outlet wire with tweezer to keep them out of contact with the specimen.

### 4. Connection and Fixation of Wires

Generally, PVC double-core multi-strand copper wire or enamel covered wire are used for connecting strain gauges and a strain indicator. The connection of lead wire and the outlet wire of the strain gauge is better transitioned by a connection terminal. The wiring terminal is fixed on the specimen with 502 glue. The wire head and the copper gauge of the wiring terminal are hung with tin in advance. Then the strain gauge's lead wire and outlet wire are welded on the terminal piece. The strain gauge's outlet wire can also be directly wound on the lead wire and then soldered with tin. Polyester insulating tape can be used to separate the solder head from the specimen. No "virtual welding" shall occur in any connection method. Finally, the pressing wire sheet (stainless steel foil) is welded to the specimen with a spot welder to fix the wire. The wire can also be fixed on the specimen with adhesive tape instead of pressing wire sheet.

## 5. Quality Inspection of Strain Gauge Sticking Process

The mounting quality is the key to the success of the electrical test, which requires not only the skilled stick techniques, but also the assurance of the appearance quality and internal quality.

(1) Appearance quality. The strain gauge mounted on the component shall have a thin and even adhesive layer and transparent sense through the sensitive gate binder. If there is too less binder and improper rolling press during mounting, bubbles will come out in sensitive gate parts, the adhesive layer will be uneven. Excessive binder will cause local uplift of the strain gauge, wrinkles of strain gauge are not allowed, and the strain gauge shall be removed and remounted. Outlet wire of the strain gauge should not be stick to the specimen.

(2) Internal quality. After the strain gauge is mounted, the resistance value will be measured with a digital multimeter. The resistance value of strain gauge should not change greatly before and after mounting. If there is a big change, it indicates that the strain gauge has been creased during mounting, and it is better to remount. After the binder is cured, the insulation resistance between the outlet wire and the component will be measured with a low voltage megohm meter. For strain gauge used for short-term measurement, the insulation resistance shall be $50\sim100$ M$\Omega$. For strain gauge used for long-term measurement, high humidity environment or underwater environment, the insulation resistance should be more than 500 M$\Omega$. The insulation resistance is an important indicator of the mounting quality of the strain gauge. As the insulation resistance is low, the zero drift, creep and hysteresis of the strain gauge are serious, which will cause large measurement errors. Insufficient curing of adhesive can also cause low insulation resistance, which can be accelerated by heating with hair dryer.

After the wire is welded, the resistance and insulation resistance should be measured again. A slight increase in the measured resistance is normal due to the resistance of the wire. However, if the reading drifts, which is generally due to bad welding, rewelding is needed. If the insulation resistance after the wire connection is found to be lower than the value before the wire connection, it is generally caused by the bottom of the wiring terminal piece being burnt through, the wiring terminal piece shall be replaced.

(3) Comprehensive quality assessment. The quality of strain gauge mounting process should be evaluated by the actual measurement. The strain indicator is a highly sensitive instrument. When the strain gauge is connected to the strain indicator, all the hidden dangers that are difficult to be found by visual inspection and multimeter measurement will be exposed. For example, bridges cannot balance due to the huge difference in the electrical resistance, drift is existed due to virtual welding or too low insulation resistance, the strain indicator changes greatly when using a rubber to press the strain gauge sensitive gate gently due to bubble and other reasons. These defects should be avoided before formal measurement.

### 6. Moisture Proof Protection for the Strain Gauge

If the mounted strain gauge is exposed to air for a long time, its bonding fastness and insulation resistance will be reduced due to dampness. If it is seriously damaged, the strain gauge will be peeled off. Therefore, a moisture proof protective layer should be laid.

Before applying the moisture proof protective layer, the coating part can be heated to 40~50 ℃ to ensure good bonding. The thickness of the protective layer is about 1~2 mm, and the surrounding area should exceed the strain gauge by 10~20 mm. You'd better bury the solder head and wiring terminals in the moisture proof protective layer.

### Ⅳ. Experiment Steps

(1) There are 4 strain gauges in each group, and 2 equal strength beam specimens.

(2) Clean the specimens.

(3) Mounting the strain gauges. After the 2 strain gauges is screened, mounting them along the longitudinal and transverse directions of the beam with equal strength according to the technological requirements, as shown in Figure 2-21.

(4) Welding. Check the welding quality and measure the electronical resistance with a multimeter, record the value of the resistance.

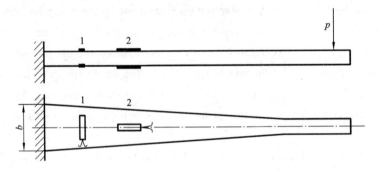

Figure 2-21   Strain Gauge Layout for a Beam with Equal Strength

### Ⅴ. Cautions

(1) The adhesive force of 502 quick-drying glue is very strong, and it has a strong pungent odor, the students should avoid excessive absorption. If the skin or clothing is stuck, it should be soaked with acetone or anhydrous alcohol, do not pull.

(2) The solder joint between the strain gauge's, outlet wire and the sensitive gate is very fragile, do not pull it out.

### Ⅵ. Questions

(1) Is it possible to omit the inspection of appearance quality and internal quality when checking the mounting quality?

(2) The longitudinal and transverse strain gauges mounted for measurement are not at-

tached to the same cross-section and are far away from the fixed end. Does this have any influence on the measurement result? If they are placed as close as possible to the fixed end, will the result be more accurate?

# Experiment 8   Experiment of the Beam with Equal Strength

## I . Objectives

(1) To be familiar with the static resistance strain indicator, and further train the bridge formation techniques in resistance strain measurement.

(2) To measure the strain on a beam with equal strength and verify the stress distribution of the beam with equal strength.

(3) To calculate the elastic modulus $E$ and Poisson's ratio $\mu$ of the beam with equal strength.

(4) To measure the weight $P$ of the object to be measured by the electrical measurement method.

## II . Experiment Equipment and Instruments

(1) An experiment device of a cantilever beam with equal strength.

(2) A static resistance strain indicator.

(3) A ruler and a vernier caliper.

## III . Experiment Content and Requirements

(1) According to the experiment purpose, think and design an experiment plan.

(2) Table designing and filling. Measuring the beam geometry ($L$, $a$, $b_0$, $b_1$ and $h$), design the strain gauge mounting scheme, bridge connection, loading, readings of the strain indicator, and etc. ,fill the corresponding tables.

(3) Calculate the elastic modulus $E$ and Poisson's ratio of the material used in the cantilever beam, and compare the materials.

(4) Measure the weight $P$ of the object to be measured.

## IV . Experiment Steps

(1) There are 4 strain gauges in each group and 2 specimens of beams with equal strength.

(2) Design mounting scheme, draw mounting diagram and data tables.

(3) After the specimen is cleaned, strain gauge mounting and welding shall be carried out according to the technical requirements of mounting.

(4) Construct an electronical bridge and connect it to the static resistance strain indicator. Load and record data.

## V. Schematic of the Beam with Equal Strength

The schematics of the beam with equal strength are shown in Figure 2-22 and Figure 2-23.

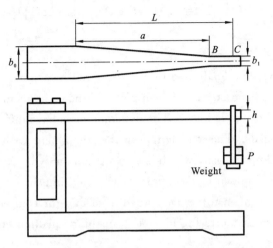

Figure 2-22  The Experiment Rig for the Beam with Equal Strength

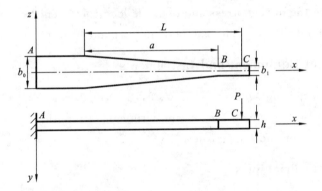

Figure 2-23  The Geometry of the Beam with Equal Strength

## VI. Questions

(1) What is a beam with equal strength?

(2) How to calculate the stress on the surface of the beam with equal strength?

(3) How to calculate the elastic modulus and Poisson's ratio of the beam with equal strength?

(4) If a 1/4-bridge or half-bridge or full-bridge is used, how should the resistance strain gauge be arranged and how should the bridges be assembled?

(5) The longitudinal and transverse strain gauges mounted for the purpose of measuring $\mu$ are not attached to the same cross-section and are far away from the fixed end, does this affect the measurement result? If you put them as close as possible to the fixed end, will the result be more accurate?

# Experiment 9   Measurement of Dynamic Coefficient of the Beam with Impact Load

The problem of dynamic loads is often encountered in engineering practice besides the static loads. The stresses and strains of each point of the component with dynamic loads are very different from that of static loads. The dynamic loads have different forms according to the different loading speeds. The load acting on the component with a great speed in a very short time is called the impact load, which is also a common dynamic load. The stress caused by the impact load when applied to a component is called impact stress. Therefore, the impact stress is the main problem to be considered in design for components that bear the impact load(such as component suffering forging, punching, drilling etc. ).

## I . Objectives

(1) To measure the impact stress and dynamic load coefficient by experiment methods.

(2) To understand the principle, method and instrument of dynamic stress measuring.

## II . Experiment Equipment and Instruments

(1) A NI data acquiring system (NIcDAQ-9178, NI9237 and NI9949).

(2) An experiment equipment for a beam with equal strength.

(3) A vernier caliper and a tape ruler.

## III . Experiment Principles

A non-uniform beam with equal strength is adopted in this experiment, as shown in Figure 2-22. At the end near the weight, it is impacted by the weight falling freely at height $H$. According to the theory, the dynamic load coefficient of the beam is

$$K_d = 1 + \sqrt{1 + \frac{2H}{\delta_j}}$$

where $H$ is the height of the weight, $\delta_j$ is the static deflection of beam. The geometry of the beam is $L = 140$ mm, $b_0 = 30$ mm, $h = 4$ mm. The beam with equal strength is shown in Figure 2-24.

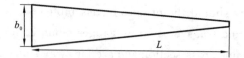

**Figure 2-24   A Simplified Model for the Beam with Equal Strength**

The weight falls at the height $H$, and impacts the end of the beam, the dynamic strain

$\varepsilon_d$ of the measured point is recorded by the NI data acquiring card. Put the weight on the same point and then the static strain $\varepsilon_j$ can be obtained. The dynamic load coefficient is

$$K_d = \varepsilon_d / \varepsilon_j$$

and the impact stress is

$$\sigma_d = E\varepsilon_d \quad \text{or} \quad \sigma_d = EK_d\varepsilon_j$$

Where $E$ is the elastic modulus (this beam is made of aluminum alloy and the elastic modulus is about 68 GPa).

## IV. Experiment Steps

(1) Measure and record the geometric dimensions, height of the weight, gravity of the weight and elastic modulus of the simply supported beam.

(2) Connecting wires. A 1/4-bridge is connected to the NI data acquiring system.

(3) Launch the self-written special software SIGNAL EXPRESS.

(4) Set parameters and calibrate the strain. Record the height of the weight, release the weight, record data and the maximum dynamic strain. The waveform diagram is shown in Figure 2-25.

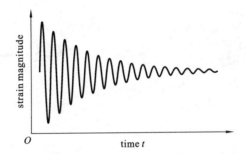

**Figure 2-25   The Diagram of Waveform**

(5) Repeat step(4) and place the weight smoothly, collect the static strain as above.

(6) Calculate the maximum impact stress and measure the dynamic load coefficient, and investigate the relationship between the coefficient and the height.

## V. Preview Requirements

(1) To review the concept and calculation method of the dynamic load coefficient.

(2) To understand dynamic strain measuring method and dynamic strain calibrating method.

## VI. Requirements for Experiment Report

The experiment report should include: the name of the experiment, the objectives of the experiment, draft of the experiment devices, name and specification of the instruments, raw data and experiment results. The experiment results should contain data recording, calculation results and curve drawing.

# Appendix

## Experiment Reports of Mechanics of Materials

### Experiment 1   Tensile Experiment of Metallic Materials

Ⅰ. **Objectives**

Ⅱ. **Experiment Equipment and Instruments**

Ⅲ. **Experiment Principles**

Ⅳ. **Experiment Steps**

## Ⅴ. Recording the Experiment Data and Calculating

Calculating accuracy:

(1) The calculation accuracy of the strength indexes (lower yield stress $\sigma_{eL}$ and tensile strength $\sigma_m$) is required to be 0.5MPa, that is, the value small than 0.25 MPa is truncated, and the value small than 0.75 MPa but equal to or bigger than 0.25 MPa is reduced to 0.5 MPa, and the value equal to or bigger than 0.75 MPa is converted to 1 MPa.

(2) The calculation accuracy of the plasticity indexes (percentage elongation after fracture $A$ and percentage reduction of area $Z$) is required to be 0.5%, that is, the value small than 0.25% is rounded off and the value small than 0.75% but equal to or bigger than 0.25% is reduced to 0.5%, and the value equal to or bigger than 0.75% is rounded to 1%.

**Table 1   Tensile Experiment Data of the Low Carbon Steel**

| | Measured data | | | 1st measurement | 2nd measurement | Average |
|---|---|---|---|---|---|---|
| Before fracture | Section 1 diameter | $d_{01}$/mm | | | | |
| | Section 2 diameter | $d_{02}$/mm | | | | |
| | Section 3 diameter | $d_{03}$/mm | | | | |
| | Original gauge length | $L_0$/mm | | | | |
| After fracture | Minimum cross-section diameter | $d_1$/mm | | | | |
| | Final gauge length after fracture | $L_u$/mm | | | | |
| | Yield load/kN | $F_{eL}$ | | | | |
| | Maximum load/kN | $F_m$ | | | | |

$\sigma_{eL} = F_{eL}/S_0 =$
$\sigma_m = F_m/S_0 =$
$A = (L_u - L_0)/L_0 \times 100\% =$
$Z = (S_0 - S_u)/S_0 \times 100\% =$

**Table 2   Tensile Experiment Data of the Gray Cast Iron**

| | Measured data | | 1st measurement | 2nd measurement | Average |
|---|---|---|---|---|---|
| Before fracture | Section 1 diameter | $d_{01}$/mm | | | |
| | Section 2 diameter | $d_{02}$/mm | | | |
| | Section 3 diameter | $d_{03}$/mm | | | |
| | Original gauge length | $L_0$/mm | | | |
| After fracture | Maximum load/kN | $F_m$ | | | |

$\sigma_m = F_m/S_0 =$

## Ⅵ. Experiment Results

Draw the tensile curves of the low carbon steel and the gray cast iron, and mark the key points.

## Ⅶ. Questions

(1) What are the similarities and differences between the mechanical properties and failure modes of the low carbon steel and the gray cast iron with normal temperature and static load?

(2) Is the percentage elongation of the specimens the same with the same diameters but different lengths? Why?

(3) What do you think are the factors causing the error to the experiment results? How can their influence be avoided or minimized?

(4) What's the practical value of measuring the mechanical properties of materials?

# Experiment 2   Torsion Experiment of Metallic Materials

### I . Objectives

### II . Experiment Equipment and Instruments

### III . Experiment Principles

### IV . Recording the Experiment Data and Calculating

Table 1   Data of the Low Carbon Steel

| Data of the low carbon steel | | 1st measurement | 2nd measurement | Average |
|---|---|---|---|---|
| Section 1 diameter | $d_{01}$/mm | | | |
| Section 2 diameter | $d_{02}$/mm | | | |
| Section 3 diameter | $d_{03}$/mm | | | |
| Yield strength | $T_{eL}$ | | | |
| Maximum torque/(N • m) | $T_m$ | | | |

Table 2   Data of the Gray Cast Iron

| Data of the gray cast iron | | 1st measurement | 2nd measurement | Average |
|---|---|---|---|---|
| Section 1 diameter | $d_{01}$/mm | | | |
| Section 2 diameter | $d_{02}$/mm | | | |
| Section 3 diameter | $d_{03}$/mm | | | |
| Maximum torque/(N • m) | $T_m$ | | | |

The low carbon steel:

$\tau_{eL} = 0.75 T_{eL}/W =$

$\tau_m = 0.75 T_m/W =$

The gray cast iron:

$\tau_m = T_m/W =$

## V. Questions

(1) Draw the torsion curves of the low carbon steel and the gray cast iron specimens, and mark the key points.

(2) Draw the fracture diagrams of the low carbon steel and the gray cast iron specimens.

(3) Compare the torsional failure fracture of the low carbon steel and the gray cast iron specimens, and analyze the failure reasons.

(4) Analyze the deformation and failure characteristics of the low carbon steel and the gray cast iron in tensile and torsion, summarize the mechanical properties.

# Experiment 3   Basic Electrical Measurement Method and the Use of Static Resistance Strain Indicator

## Ⅰ. Experiment Principles

（1）What is the electrical measurement method? What are the advantages and disadvantages of this method?

（2）What is the resistance strain effect?

（3）What is a strain bridge? Describe the principles of strain bridge briefly.

（4）How many ways can strain gauge be connected in a bridge? Draw the corresponding bridge schematic diagram.

（5）What is the temperature effect? What are the measures to eliminate the temperature effect?

## Ⅱ. Experiment Steps

Strain gauge connection experiment: Taking the bending normal stress experiment device of the beam with uniform section as an example, the single-arm connection method of half-bridge (1/4-bridge), double-arm connection method of half-bridge and full-bridge connection method are practiced. Please draw the bridge wiring diagram of the strain gauge and write down the strain formula.

(1) Single-arm connection method of half-bridge.

(2) Double-arm connection method of half-bridge.

(3) Full-bridge connection method.

# Experiment 4   Measurement Experiment of Bending Normal Stress of Rectangular Beam (1/4-bridge)

## Ⅰ. Objectives

## Ⅱ. Experiment Equipment and Instruments

## Ⅲ. Experiment Principles

## Ⅳ. Experiment Steps

## Ⅴ. Recording the Experiment Data and Calculating

Table 1   Geometry of the Rectangular Beam

| $a$/mm | $b$/mm | $c$/mm | $L$/mm | $h$/mm | $E$/GPa | $K$ | $R_0$ |
|--------|--------|--------|--------|--------|---------|-----|-------|
|        |        |        |        |        |         |     |       |

**Table 2    Data of the Rectangular Beam in the Bending Experiment**

| Load/kN | Strain($\mu\varepsilon$) | | | | | | | |
|---------|-------------|-------------|--------------|-------------|----------------|----------------|-------------|--------------|
| | $y_1 = 0$ | $y_2 = 0$ | $y_3 = h/4$ | $y_4 = h/4$ | $y_5 = -h/4$ | $y_6 = -h/4$ | $y_7 = h/2$ | $y_8 = -h/2$ |
| 1 | | | | | | | | |
| 1.5 | | | | | | | | |
| 2 | | | | | | | | |
| 2.5 | | | | | | | | |
| 3 | | | | | | | | |
| $\overline{\varepsilon_i} =$ | $\overline{\varepsilon_1} =$ | $\overline{\varepsilon_2} =$ | $\overline{\varepsilon_3} =$ | $\overline{\varepsilon_4} =$ | $\overline{\varepsilon_5} =$ | $\overline{\varepsilon_6} =$ | $\overline{\varepsilon_7} =$ | $\overline{\varepsilon_8} =$ |
| $\sigma_{real} = E\overline{\varepsilon}$ | | | | | | | | |
| $\sigma_{theory} = \dfrac{\overline{M}y}{I_z}$ | | | | | | | | |

(1) $\overline{\varepsilon_0} =$

$\sigma_{real1} = E\overline{\varepsilon_1} =$

$\sigma_{theory1} = \dfrac{\overline{M}y}{I_z} =$

Relative error (%):

(2) $\overline{\varepsilon_2} =$

$\sigma_{real2} = E\overline{\varepsilon_2} =$

$\sigma_{theory2} = \dfrac{\overline{M}y}{I_z} =$

Relative error (%):

(3) $\overline{\varepsilon_3} =$

$\sigma_{real3} = E\overline{\varepsilon_3} =$

$\sigma_{theory3} = \dfrac{\overline{M}y}{I_z} =$

Relative error (%):

(4) $\overline{\varepsilon_4} =$

$\sigma_{real4} = E\overline{\varepsilon_4} =$

$\sigma_{theory4} = \dfrac{\overline{M}y}{I_z} =$

Relative error (%):

(5) $\overline{\varepsilon_5} =$

$\sigma_{real5} = E\overline{\varepsilon_5} =$

$\sigma_{theory5} = \dfrac{\overline{M}y}{I_z} =$

Relative error (%):

(6) $\overline{\varepsilon_6} =$

$\sigma_{real6} = E\overline{\varepsilon_6} =$

$$\sigma_{\text{theory6}} = \frac{\overline{M}y}{I_z} =$$

Relative error (%):

(7) $\overline{\varepsilon_7} =$

$$\sigma_{\text{real7}} = E\overline{\varepsilon_7} =$$

$$\sigma_{\text{theory7}} = \frac{\overline{M}y}{I_z} =$$

Relative error (%):

(8) $\overline{\varepsilon_8} =$

$$\sigma_{\text{real8}} = E\overline{\varepsilon_8} =$$

$$\sigma_{\text{theory8}} = \frac{\overline{M}y}{I_z} =$$

Relative error (%):

### Table 3  Data of the Rectangular Beam in the Transverse Bending Experiment

| Load/kN | Strain($\mu\varepsilon$) | | | | | | | |
|---|---|---|---|---|---|---|---|---|
| | $y_1 = 0$ | $y_2 = 0$ | $y_3 = h/4$ | $y_4 = h/4$ | $y_5 = -h/4$ | $y_6 = -h/4$ | $y_7 = h/2$ | $y_8 = -h/2$ |
| 1 | | | | | | | | |
| 1.5 | | | | | | | | |
| 2 | | | | | | | | |
| 2.5 | | | | | | | | |
| 3 | | | | | | | | |
| $\overline{\varepsilon_i} =$ | $\overline{\varepsilon_1} =$ | $\overline{\varepsilon_2} =$ | $\overline{\varepsilon_3} =$ | $\overline{\varepsilon_4} =$ | $\overline{\varepsilon_5} =$ | $\overline{\varepsilon_6} =$ | $\overline{\varepsilon_7} =$ | $\overline{\varepsilon_8} =$ |
| $\sigma_{\text{real}} = E\overline{\varepsilon}$ | | | | | | | | |
| $\sigma_{\text{theory}} = \dfrac{\overline{M}y}{I_z}$ | | | | | | | | |

(1) $\overline{\varepsilon_1} =$

$$\sigma_{\text{real1}} = E\overline{\varepsilon_1} =$$

$$\sigma_{\text{theory1}} = \frac{\overline{M}y}{I_z} =$$

Relative error (%):

(2) $\overline{\varepsilon_2} =$

$$\sigma_{\text{real2}} = E\overline{\varepsilon_2} =$$

$$\sigma_{\text{theory2}} = \frac{\overline{M}y}{I_z} =$$

Relative error (%):

(3) $\overline{\varepsilon_3} =$

$$\sigma_{\text{real3}} = E\overline{\varepsilon_3} =$$

$$\sigma_{\text{theory3}} = \frac{\overline{M}y}{I_z} =$$

Relative error (%):

(4) $\overline{\varepsilon_4} =$

$\sigma_{real4} = E\overline{\varepsilon_4} =$

$\sigma_{theory4} = \dfrac{\overline{M}y}{I_z} =$

Relative error (%):

(5) $\overline{\varepsilon_5} =$

$\sigma_{real5} = E\overline{\varepsilon_5} =$

$\sigma_{theory5} = \dfrac{\overline{M}y}{I_z} =$

Relative error (%):

(6) $\overline{\varepsilon_6} =$

$\sigma_{real6} = E\overline{\varepsilon_6} =$

$\sigma_{theory6} = \dfrac{\overline{M}y}{I_z} =$

Relative error (%):

(7) $\overline{\varepsilon_7} =$

$\sigma_{real7} = E\overline{\varepsilon_7} =$

$\sigma_{theory7} = \dfrac{\overline{M}y}{I_z} =$

Relative error (%):

(8) $\overline{\varepsilon_8} =$

$\sigma_{real8} = E\overline{\varepsilon_8} =$

$\sigma_{theory8} = \dfrac{\overline{M}y}{I_z} =$

Relative error (%):

## Ⅵ. Questions

(1) If the load is fixed to be a constant, how is the strain distributed along the height of the beam? What is the theoretical basis? Write down the formula.

(2) At the same measuring point, how does the strain change as the increase of the load at the same measuring point? What is the theoretical basis? Write down the formula.

# Experiment 5   Measurement Experiment of Bending Normal Stress of Rectangular Beam ( Half - bridge and Full - bridge )

## Ⅰ. Objectives

## Ⅱ. Experiment Equipment and Instruments

## Ⅲ. Experiment Principles

## Ⅳ. Experiment Steps

## V. Recording the Experiment Data and Calculating

Table 1    Geometry of the Beam

| $a$/mm | $b$/mm | $c$/mm | $L$/mm | $h$/mm | $E$/GPa | $K$ | $R_0$ |
|---|---|---|---|---|---|---|---|
| | | | | | | | |

Table 2    Data of Half-bridge Method

| Load/kN | Strain($\mu\varepsilon$) | | | | | |
|---|---|---|---|---|---|---|
| | Pure bending | | | Transverse bending | | |
| | $y_1 =$ | $y_2 =$ | $y_3 =$ | $y_4 =$ | $y_5 =$ | $y_6 =$ |
| 1 | | | | | | |
| 1.5 | | | | | | |
| 2 | | | | | | |
| 2.5 | | | | | | |
| 3 | | | | | | |
| $\overline{\varepsilon_i} =$ | $\overline{\varepsilon_1} =$ | $\overline{\varepsilon_2} =$ | $\overline{\varepsilon_3} =$ | $\overline{\varepsilon_4} =$ | $\overline{\varepsilon_5} =$ | $\overline{\varepsilon_6} =$ |
| $\sigma_{real} = \dfrac{E\overline{\varepsilon}}{2}$ | | | | | | |
| $\sigma_{theory} = \dfrac{\overline{M}y}{I_z}$ | | | | | | |
| Relative error | | | | | | |

Pure bending:

(1) $\overline{\varepsilon_1} =$

$\sigma_{real1} = E\overline{\varepsilon_1}/2 =$

$\sigma_{theory1} = \dfrac{\overline{M}y}{I_z} =$

Relative error (%):

(2) $\overline{\varepsilon} =$

$\sigma_{real2} = E\overline{\varepsilon_2}/2 =$

$\sigma_{theory2} = \dfrac{\overline{M}y}{I_z} =$

Relative error (%):

(3) $\overline{\varepsilon_3} =$

$\sigma_{real3} = E\overline{\varepsilon_3}/2 =$

$\sigma_{theory3} = \dfrac{\overline{M}y}{I_z} =$

Relative error (%):

Transverse bending:

(4) $\overline{\varepsilon_4} =$

$\sigma_{real4} = E\overline{\varepsilon}_4/2 =$

$\sigma_{theory4} = \dfrac{\overline{M}y}{I_z} =$

Relative error (%):

(5) $\overline{\varepsilon}_5 =$

$\sigma_{real5} = E\overline{\varepsilon}_5/2 =$

$\sigma_{theory5} = \dfrac{\overline{M}y}{I_z} =$

Relative error (%):

(6) $\overline{\varepsilon}_6 =$

$\sigma_{real6} = E\overline{\varepsilon}_6/2 =$

$\sigma_{theory6} = \dfrac{\overline{M}y}{I_z} =$

Relative error (%):

**Table 3   Data of the Full-bridge Method**

| Load/kN | Strain($\mu\varepsilon$) | |
|---|---|---|
| | Pure bending | Transverse bending |
| | $y_1 =$ | $y_2 =$ |
| 1 | | |
| 1.5 | | |
| 2 | | |
| 2.5 | | |
| 3 | | |
| $\overline{\varepsilon}_i =$ | $\overline{\varepsilon}_1 =$ | $\overline{\varepsilon}_2 =$ |
| $\sigma_{real} = E\overline{\varepsilon}/4$ | | |
| $\sigma_{theory} = \dfrac{\overline{M}y}{I_z}$ | | |
| Relative error (%) | | |

Pure bending:

(1) $\overline{\varepsilon}_1 =$

$\sigma_{real1} = E\overline{\varepsilon}_1/4 =$

$\sigma_{theory1} = \dfrac{\overline{M}y}{I_z} =$

Relative error (%):

Transverse bending:

(2) $\overline{\varepsilon}_2 =$

$\sigma_{real2} = E\overline{\varepsilon}_2/4 =$

$\sigma_{theory2} = \dfrac{\overline{M}y}{I_z} =$

Relative error (%):

## Ⅵ. Questions

(1) What may be the reasons that the measured strain in the neutral layer is not zero? Try to analyze the strain near the measuring point to explain.

(2) Analyze the similarities and differences between the half-bridge, full-bridge and 1/4-bridge methods in this experiment.

(3) What are the main factors that affect the result of the experiment?

(4) Is the normal stress in bending affected by the elastic modulus of the material? Why?

# Experiment 6　Determination of Principal Stress Direction in Bending and Torsion Combination

## Ⅰ. Objectives

## Ⅱ. Experiment Equipment and Instruments

## Ⅲ. Experiment Principles

(1) Draw the simplified structure diagram of the experiment device and the free-body diagram.

(2) Introduce the principles of strain analysis for the plane stress state.

## IV. Recording the Experiment Data and Calculating

**Table 1  Geometries of the Thin-wall Cylinder and the Parameters of Strain Rosette**

| Outer diameter of the thin-wall cylinder $D$/mm | Inner diameter of the thin-wall cylinder $d$/mm | $L$/mm | $a$/mm | Poisson's ratio $\mu$ | Elastic modulus $E$/GPa | Sensitivity coefficient $K$ | Resistance $R_0$ |
|---|---|---|---|---|---|---|---|
|  |  |  |  |  |  |  |  |

Note: The actual sizes of thin-wall cylinders after processing are different, the data are labeled on the instrument of each group.

**Table 2  Strains Measured for Points $A$ and $B$**

| Load/N | Strain ($\mu\varepsilon$) | | | | | |
|---|---|---|---|---|---|---|
|  | Upper point $A$ | | | Lower point $B$ | | |
|  | $\varepsilon_{-45°}$ | $\varepsilon_{0°}$ | $\varepsilon_{45°}$ | $\varepsilon_{-45°}$ | $\varepsilon_{0°}$ | $\varepsilon_{45°}$ |
| 9.8 |  |  |  |  |  |  |
| $2 \times 9.8$ |  |  |  |  |  |  |
| $3 \times 9.8$ |  |  |  |  |  |  |
| $4 \times 9.8$ |  |  |  |  |  |  |
| $5 \times 9.8$ |  |  |  |  |  |  |
| $\overline{\varepsilon_i} =$ | $\overline{\varepsilon_{-45°}} =$ | $\overline{\varepsilon_{0°}} =$ | $\overline{\varepsilon_{45°}} =$ | $\overline{\varepsilon_{-45°}} =$ | $\overline{\varepsilon_{0°}} =$ | $\overline{\varepsilon_{45°}} =$ |

(1) Upper point $A$:

$\overline{\varepsilon_{-45°}} =$

$\overline{\varepsilon_{0°}} =$

$\overline{\varepsilon_{45°}} =$

Principal direction:

$$\tan 2\alpha_0 = \frac{\varepsilon_{45°} - \varepsilon_{-45°}}{2\varepsilon_{0°} - \varepsilon_{-45°} - \varepsilon_{45°}} =$$

$\alpha_0 =$

Principal strains: $\varepsilon_{1,3} = \dfrac{\varepsilon_{45°} + \varepsilon_{-45°}}{2} \pm \dfrac{\sqrt{2}}{2} \sqrt{(\varepsilon_{-45°} - \varepsilon_{0°})^2 + (\varepsilon_{45°} - \varepsilon_{0°})^2} =$

$\varepsilon_1 =$

$\varepsilon_3 =$

Principal stresses: $\sigma_1 = \dfrac{E}{1 - \mu^2}(\varepsilon_1 + \mu\varepsilon_3) =$

$\sigma_3 = \dfrac{E}{1 - \mu^2}(\varepsilon_3 + \mu\varepsilon_1) =$

(2) Lower point $B$:

$\overline{\varepsilon_{-45°}} =$

$\overline{\varepsilon_{0°}} =$

$\overline{\varepsilon_{45°}} =$

Principal direction: $\tan 2\alpha_0 = \dfrac{\varepsilon_{45°} - \varepsilon_{-45°}}{2\varepsilon_{0°} - \varepsilon_{-45°} - \varepsilon_{45°}} =$

$\alpha_0 =$

Principal strains: $\varepsilon_{1,3} = \dfrac{\varepsilon_{45°} + \varepsilon_{-45°}}{2} \pm \dfrac{\sqrt{2}}{2} \sqrt{(\varepsilon_{-45} - \varepsilon_{0°})^2 + (\varepsilon_{45°} - \varepsilon_{0°})^2} =$

$\varepsilon_1 =$

$\varepsilon_3 =$

Principal stresses: $\sigma_1 = \dfrac{E}{1-\mu^2}(\varepsilon_1 + \mu\varepsilon_3) =$

$\sigma_3 = \dfrac{E}{1-\mu^2}(\varepsilon_3 + \mu\varepsilon_1) =$

## Ⅴ. Questions

(1) Are the principal stresses and directions the same for point $A$ and point $B$? Why? Please draw a diagram to explain.

(2) Calculate the theoretical principal stresses and directions for point $A$ and calculate the error between the results with those obtained from the experiment. Analyze the factors that lead to the error.

# Experiment 7   Strain Gauges Mounting

## Ⅰ. Objectives

## Ⅱ. Experiment Equipment and Instruments

## Ⅲ. Experiment Principles

## Ⅳ. Experiment Steps

(1) Draw the experimental device diagram and mounting diagram.

(2) Measure the electrical resistance before and after the strain gauge is mounted.

(3) How to guarantee the mounting quality of the strain gauge mounted?

# Experiment 8 Experiment of the Beam with Equal Strength

## I. Objectives

## II. Experiment Equipment and Instruments

## III. Experiment Plan

(1) Write down the formulas of the elastic modulus $E$ and Poisson's ratio $\mu$ of a beam with equal strength.

(2) Design the strain bridge.

(3) Draw the gauge mounting diagram.

## IV. Recording the Experiment Data and Calculating

(1) Draw the geometrical diagram of the beam with equal strength.

(2) Design data tables and record experiment data.

(3) Calculate $E$ and $\mu$ of the beam with equal strength, judge the material of the beam with equal strength by checking related file.

(4) Measure the weight of the object by the electrical measurement method.

# Experiment 9　Measurement of Dynamic Coefficient of Beams with Impact Load

Ⅰ. **Objectives**

Ⅱ. **Experiment Equipment and Instruments**

Ⅲ. **Experiment Principle**

Ⅳ. **Sketch of the Experiment Unit**

# V. Experiment Steps

# VI. Recording Experiment Results

# VII. Experiment Analysis

# 参考文献(References)

[1] 王杏根.工程力学实验[M].武汉:华中科技大学出版社,2008.

[2] 张亦良.工程力学实验[M].北京:北京工业大学出版社,2010.

[3] 范钦珊.工程力学实验指导教程[M].南京:南京大学出版社,2010.

[4] 董雪花.工程力学实验[M].北京:国防工业出版社,2011.

[5] 邓宗白.材料力学实验与训练[M].北京:高等教育出版社,2014.

[6] 刘鸿文.材料力学实验[M].北京:高等教育出版社,2017.

[7] 束德林.工程材料力学性能[M].北京:机械工业出版社,2017.

[8] 杨绪普.工程力学实验[M].北京:中国铁道出版社,2018.

[9] 中国钢铁工业协会.金属材料 拉伸试验 第1部分:室温试验方法:GB/T 228.1—2010[S].北京:中国标准出版社,2010.

[10] 中国钢铁工业协会.金属材料 室温扭转试验方法:GB/T 10128—2007[S].北京:中国标准出版社,2007.

[11] 中国机械工业联合会.金属粘贴式电阻应变计:GB/T 13992—2010[S].北京:中国标准出版社,2010.

[12] HIBBELER R C. Mechanics of Materials[M]. 8th edition. New Jersey:Pearson Prentice Hall,2011.